Minds Make Societies

Minds Make Societies

How Cognition Explains the World Humans Create

Pascal Boyer

Yale

UNIVERSITY PRESS

New Haven and London

Published with assistance from the foundation established in memory of Philip Hamilton McMillan of the Class of 1894, Yale College.

Yale University Press books may be purchased in quantity for educational, business, or promotional use. For information, please e-mail sales.press@yale.edu (U.S. office) or sales@yaleup.co.uk (U.K. office).

Set in Bulmer type by IDS Infotech, Ltd.
Printed in the United States of America.

Library of Congress Control Number: 2017958313
ISBN 978-0-300-22345-3 (hardcover : alk. paper)

A catalogue record for this book is available from the British Library.

This paper meets the requirements of ANSI/NISO Z39.48-1992 (Permanence of Paper).

10 9 8 7 6 5 4 3 2 1

To Louis and Claire, from afar

Contents

SIX
Can Human Minds Understand Societies?
Coordination, Folk Sociology, and Natural Politics   203

CONCLUSION:
COGNITION AND COMMUNICATION CREATE
TRADITIONS   245

Preface

HUMANKIND IS SPECIAL IN MANY WAYS, as the result of its evolutionary history. Among other unique features, humans are special in that they build complex and apparently very different societies. Recent developments in different sciences are now converging to provide explanations for many aspects of these societies, for the particular ways in which humans for instance create hierarchies, families, gender norms, economic systems, group conflict, moral norms, and much, much more. This exciting scientific development, still in its early age, is what I describe in this book.

This new perspective on human societies did not originate in a flash of inspiration, in the revelation of a new theory of societies. That is not the way sciences work. Rather, what I present here is an accumulation of very specific scientific findings, in various fields like evolutionary biology, cognitive psychology, archaeology, anthropology, economics, and more. Rather than giving us a general theory of society, this perspective offers specific answers to specific questions, such as, Why do people want a just society? Is there a natural form of the family? What makes men and women behave differently? Why are there religions? Why do people participate in conflicts between groups? And so forth.

That is why each chapter in the main part of this book deals with one of these crucial questions. In each chapter, I explain how one of these crucial questions is addressed in a new and surprising way in this new kind of social science, which of course does not amount to providing the definitive answers to any of the questions in the titles. Nor do these chapters offer a survey of all that has been written about these various questions—in each case a long book would barely suffice.

Most of the findings and ideas presented here, and probably all the important ones, are other people's. I merely provide the connections between these elements of a new approach to human societies. I have included multiple references, not just to give credit where due, but also to orient curious readers to the best in recent scholarship. The endnotes only provide references to these sources. They contain no other material, to avoid distracting from the argument.

Minds Make Societies

Introduction

Human Societies through the Lens of Nature

WHY SHOULD SOCIETY BE A MYSTERY? There is no good reason human societies should not be described and explained with the same precision and success as the rest of nature. And there is every reason to hope that we can understand social processes, as their impact on our lives is so great. Since there is no better way than science to understand the world, surely a science of what happens in human societies is devoutly to be wished.

But until recently we had nothing of the kind. This was not for lack of effort. For centuries, students of societies had collected relevant facts about different societies. They had tried to compare places and times and make sense of it all, often desultorily groping for principles of society or history that would emulate the clarity of natural laws. In many cases this effort proved fascinating and illuminating, from Ibn Khaldūn and Montesquieu to Tocqueville, Adam Smith, and Max Weber—and many others. But there was little sense of cumulative progress.

All this is changing, mostly because evolutionary biology, genetics, psychology, economics, and other fields are converging to propose a unified understanding of human behavior. Over the past few decades, a variety of scientific fields have made great progress in explaining some crucial parts of what makes humans special—in particular, how humans build and organize societies. The main reason for this progress was a radical shift from tradition. The social sciences, at some point in their history, had made the disastrous mistake of considering human psychology and evolution of no importance. The idea was that understanding history and society would not

1

require much knowledge of how humans evolved and how their organs function. In that view, the natural sciences could tell you many things of great interest about humans, about the reasons we have lungs and hearts, about the way we digest or reproduce—but they could never explain why people would storm the Winter Palace or throw chests of tea into Boston Harbor.

But that was all wrong. As it happens, findings from evolutionary biology and psychology as well as other empirical sciences are crucial to explaining such events, and social processes in general. In the past fifty years or so, the scientific study of humankind has made great strides, producing ever-greater knowledge of how brains work and how evolution shaped organisms, as well as formal models of how people interact and how such local interactions give rise to global dynamics.

That we would progress toward this more unified perspective on human societies was anticipated for some time.[1] But it is only recently that the study of how people form and manage groups has been turning into a proper scientific enterprise, with many difficult questions and frustrating uncertainties—but also surprisingly clear results.

What Sort of Things Do We Want Explained?

One should never start with theory. Instead of first principles and deductions, let me offer a collage—a ragtag, fragmentary, and unorganized list of phenomena we would want a proper social science to explain.

WHY DO PEOPLE BELIEVE SO MANY THINGS THAT AIN'T SO?

All over the world, a great many people seem to believe things that others judge clearly absurd. The repertoire of what counts as reasonable in one place, and utter nonsense in another, is vast. Some people fear that contact with outsiders will make their penis disappear, while others hope that reciting a formula can make a stranger fall in love with them. People transmit to each other all kinds of rumors and urban legends. Some say that the AIDS epidemic was engineered by the secret services. Others maintain that the machinations of witches are certainly the explanation for illness and misfortune. It would seem that human minds are exceedingly vulnerable to low-quality information—and that scientific or technical progress seems to make little difference.

WHY POLITICAL DOMINATION?

Man, some have said, is born free yet everywhere is in chains. Why do human beings tolerate domination? Social scientists, it seems, should try to explain to us how political domination can emerge and subsist in human groups. They should explain people's submission to autocratic emperors for most of Chinese history, their enthusiasm for nationalistic demagoguery in twentieth-century Europe, their cowed acceptance of totalitarian communist regimes for seventy years, or their toleration of kleptocratic dictators in many parts of contemporary Africa. If it is true that the history of most hitherto existing society is the history of domination by kings, warlords, and elites, what makes such oppression possible, and durable?

WHY ARE PEOPLE SO INTERESTED IN ETHNIC IDENTITY?

All over the world, and for as long as records exist, people have considered themselves members of groups, most often of ethnic groups, that is, of supposedly common descent. People readily construe the world as a zero-sum game between their own and other ethnic groups, which justifies all manner of segregation and discrimination, and easily leads to ethnic strife or even warfare. Why do people find such ideas compelling and seem prepared to incur large costs in the pursuit of ethnic rivalry?

WHAT MAKES MEN AND WOMEN DIFFERENT?

In all human societies there are distinct gender roles, that is, common expectations about the way women and men typically behave. Where do these come from? How do they relate to differences in anatomy and physiology?

Also, if there are distinct gender roles, why are they so often associated with differences in influence and power?

ARE THERE DIFFERENT POSSIBLE MODELS OF THE FAMILY?

Related to gender roles, there are considerable debates in modern societies about the proper or natural form of the family. Is there such a thing? Children require parents, but how many and which ones, and in what arrangements? These discussions are often conducted in terms of ideology rather than appeal to scientific facts. But what are the scientific facts about the diversity and common features of human families? Do these facts tell us what forms of the family are more viable, or what problems beset them?

WHY ARE HUMANS SO UNCOOPERATIVE?

Humans spend a great deal of energy in conflict, between individuals and between groups. The frequency and nature of conflicts, and the extent to which they lead to violence, vary a lot between places. What explains such differences? Also, is human conflict an inevitable consequence of our nature? For instance, people used to think that there was some aggressive urge in humans that needed to be released, a bit like pressure building up inside a furnace until the steam escapes through a safety valve. Is that a plausible description of human motivations? If not, what explains violence and aggression?

WHY ARE HUMANS SO COOPERATIVE?

The obverse of conflict is cooperation, which attracts less attention, probably because it is ubiquitous and therefore invisible. Humans are extraordinarily cooperative. They routinely engage in collective action, in which people coordinate their actions to get better results than they would in isolation. People in small-scale societies go hunting or gathering food together, and they often share most of the proceeds. In modern societies they join associations or political parties to achieve particular goals. Is there a cooperative instinct in human beings? If so, what conditions favor or hinder its expression?

COULD SOCIETY BE JUST?

In most human societies there are class or rank distinctions, and production results in unequal incomes and wealth. In some cases the difference seems a simple effect of political dominance. Warlords, aristocrats, dictators, or the *nomenklatura* of communist regimes simply appropriate the best resources. But in most modern democracies the economic process leads to unequal outcomes without such direct theft. The main question of modern politics is, What to do with such outcomes? But this question itself raises many others that our social scientists should be able to answer, for example, What do people mean when they say they want a just society? Why does that goal motivate people to advocate diametrically opposite policies? Is there a common human notion of justice, or does it differ from place to place? Can humans actually understand the complex processes that lead to unjust or unequal outcomes?

WHAT EXPLAINS MORALITY?

Why do we have moral feelings and strong emotional reactions to violations of moral norms? People the world over have moral norms and pass moral judgment, but do they do so on the basis of the same values? And how does morality enter the minds of young children? Many moralists described human nature as entirely amoral, suggesting that ethical feelings and motivations were somehow planted in our minds by "society." But how would that happen?

WHY ARE THERE RELIGIONS?

There are organized religions in many places in the world. In small-scale societies there are no religious organizations, but people talk about spirits and ancestors. So it seems that humans have a general susceptibility to such notions. Is there a religious instinct, some specific part of the mind that creates these ideas about supernatural powers and agents, those gods and spirits? Or, on the contrary, do these religious representations illustrate some possible dysfunction of the mind? In either case, how do we explain that religious activities include collective events? How do we explain that humans seem to entertain such an extraordinary variety of religious ideas?

WHY DO PEOPLE MONITOR AND REGIMENT OTHER PEOPLE'S BEHAVIORS?

The world over, people seem to be greatly interested in moralizing, regulating, and generally monitoring other people's behaviors. This is of course very much the case in small-scale groups, where one lives under the tyranny of the cousins, as some anthropologists described it. But in large, modern societies, we also see that people are greatly interested in others' mores, sexual preferences, the way they marry or what drugs they take. This certainly goes beyond self-interest and raises the question, Is it part of human nature to meddle?

There is no particular order in this disparate list, and no coherence either—some of these are very broad questions, and others much more specific. Some are questions that many philosophers and writers have labored on over centuries, others only occur if you know human prehistory. People from another time or another place would certainly ask different questions, or would formulate them very differently. The point here is to suggest the kind of questions that we expect social scientists to address.

These questions are at the center of many contemporary debates throughout the modern world, about, for example, the role of ethnic identity in making nations, the effects of economic systems on social justice, possible models of the family and gender relations, the dangers posed by extremist religions, or the consequences of what is often called the information revolution. Because these are pressing and important questions, there is of course a great demand for easy, sweeping answers that would both describe and prescribe, tell us how society works and by the same token how to make it better. Many political ideologies are based on that kind of promise of a magic bullet that answers most issues and provides a guide for action as well, a solution that is neat, plausible, and wrong, as H. L. Mencken put it.

But we can do better. In particular, we can step back and ask, What do we actually know about the human dispositions, capacities, or preferences involved in all these behaviors? One may be surprised by how much we already know about, for example, the way a human mind acquires beliefs and evaluates their plausibility, how humans become attached to their groups or tribes and conceive of other groups, what motivations are involved in building and maintaining families, or about the processes that make women and men so different and similar.[2]

Rule I: See the Strangeness of the Familiar

Springboks sometimes jump up in the air with great élan (they "stot" or "pronk") when they detect a lion, an apparently self-defeating course of action, as it makes them far more noticeable to the lurking predator. Peacocks sport a large and apparently useless train of beautiful long feathers. Such traits and behaviors appear surprising enough. It seems to us that gazelles should know better than to attract a predator's attention, and that peacocks are just wasting energy carrying around such heavy pageantry.

An extraterrestrial anthropologist might well feel, and a proper social scientist should certainly feel, equally baffled by many aspects of human behavior. Why do humans form stable groups, in competition and occasional conflict with other groups? Why do they show any attachment to their groups, sometimes at the expense of their own welfare? Why do they imagine deities and engage in religious ceremonies? Why do men and women form stable unions and jointly nurture their children? Why are people apparently concerned with justice and inequality? And so forth.

Cultural anthropologists used to encourage their students to investigate the norms and practices of people in distant places, on the sensible assumption that familiarity is one of the biggest obstacles on the way to understanding social phenomena. Sacrificing a bull to placate invisible ancestors seems very sensible to many East African pastoralists. Having an entire organization, with specialized personnel and buildings, dedicated to managing relationships with invisible deities seems just as reasonable to many Christians or Muslims. Only unfamiliar customs prompt us to seek explanations. It was a French aristocrat, after all, who wrote the best description of the early American republic. Tocqueville was familiar with ancien régime absolutism and the revolutionary Terror—which is precisely why he could see American democracy as a perplexing oddity, in need of an explanation.

Understanding very general features of human cultures, like the existence of marriage or religious beliefs or moral feelings, requires the same estrangement procedure. Only this time we must step aside, not just from local norms, but from humanity itself. How could we do that? The economist Paul Seabright suggested that we should consider human behavior from the viewpoint of other animals. If bonobos, for instance, studied humans, they would marvel at the way we spend an inordinate amount of time and energy thinking about sex, longing for it, imagining it, singing, talking, and writing about it, while actually doing it so very rarely—rarely, that is, compared to bonobos. Gorillas would be astonished that the leader in human groups is not always the most formidable individual, and puzzle over the question of how weaklings manage to exert authority over the big bruisers. A chimpanzee anthropologist would wonder how humans can huddle together in large crowds without constant fights, why they often remain attached to the same sexual partner for years, and why fathers are at all interested in their offspring.[3]

Fortunately, we do not need to adopt a gorilla's or a chimpanzee's viewpoint. We can look at human behaviors from the outside, because evolutionary biology allows us to do so. In fact it requires us to do so and, together with other scientific disciplines, provides the necessary tools. The evolutionary perspective assumes that we inherited specific human capacities and dispositions, different from those of other organisms, because they contributed to the fitness, the reproductive potential, of our ancestors. Seen from that evolutionary standpoint, all human cultures do seem exceedingly strange, all customs seem to cry out for explanation. Most of what humans do, like form groups and have marriages and pay attention to their offspring and imagine supernatural beings, becomes slightly mysterious. It could have been otherwise. And in most animal species it certainly is. As the anthropologist Rob Foley put it, only evolution can explain those many ways in which we are just another unique species.[4]

The evolutionary perspective also allows us to go past all the easy, shot-from-the-hip answers that crop up all too often in our spontaneous reflections about human behaviors, for example, Why do people want to have sex? Because it is pleasurable. Why do humans crave sugar and fat? Because they taste good. Why do we abhor vomit? Because it smells terrible. Why do most people seek the company of funny people? Because laughing is pleasant. Why are people often xenophobic? Because they prefer their own ways and customs to those of others.

In evolutionary terms, these explanations of course have it back to front. We do not like sugar because it tastes good and abhor vomit because it is foul smelling. Rather, one is delicious and the other repulsive because we were designed to seek the former and avoid the latter. We evolved in environments in which sugar was rare enough that taking all you could was a good strategy, and vomit was certainly full of toxins and pathogens. Individuals who showed these preferences, a bit more than others, would extract more calories and fewer dangerous substances from their environments. On average, these individuals would have an ever so slightly better chance of having offspring than those others. In more accurate terms—some genes provided organisms with a moderate interest in ripe fruit, while other variants prompted a keener motivation to eat them and greater pleasure in their consumption. The latter genes gradually became more frequent in human populations. That is all we mean when we say that a strong craving for sugar is an evolved property of today's humans, and the same goes for the avoidance of regurgitated food. This evolutionary logic, so easy to understand and even easier to misunderstand, is a key to explaining human behaviors, including the way we live in societies.

Fine, one might think, our evolved nature could explain the fact that we live in societies, but could it explain the different ways we live, in different societies with different norms? After all, the questions I listed above are all about processes that differ from place to place. Family relations are different in Iceland, Japan, and the Congo. The way we commonly frame social justice issues is certainly different in modern mass societies, agrarian kingdoms, and small groups of foragers. Religious doctrines and beliefs in magic also seem very different in different places. So, could a study of human evolved nature, presumably the same in all these places, explain such diverse outcomes?

To answer this question would (and in fact does) take a whole book. But the main answer is that, yes, our evolved capacities and dispositions do explain the way we live in societies, and many important differences between times and places. But we cannot, and should not try to, demonstrate that in theoretical terms. Rather, we can examine some important domains, like the form of human families and the existence of political dominance, and see how they make much more sense once we know more about the human dispositions involved.

GETTING INFORMATION FROM ENVIRONMENTS

Let us step back. To understand the logic of evolutionary explanations, including the explanation of complex social behaviors, we must describe the way organisms in general pick up information from environments. Rather than laboring the point in theoretical terms, it may be more help to start with an example.

In many species of birds, reproduction follows the seasons. Beginning in spring, males and females size each other up, select an attractive partner, build a nest, mate, and are blessed with a few eggs that promptly hatch. The parents feed their offspring for several weeks, after which they all part. At the beginning of autumn, the cycle could in principle start again, but the birds now seem to have lost the appetite for sex and parenting. This makes sense, as food is most abundant precisely when needed to feed the offspring. In migratory species, late summer and early autumn is also a time when individuals regrow feathers and build up muscle mass in preparation for long journeys. So they need to sustain themselves rather than bring worms to hungry squawkers. In many species the sex organs shrink during that season.[5]

This yearly schedule of reproduction constitutes an adaptation to the ecology of middle and high latitudes of Eurasia and the Americas. The environment simply could not support more than one clutch a year; the time required for courtship, nest building, mating, and feeding offspring demands that one start early in spring. Once-a-year reproduction is optimal, given these conditions. It is an evolved property of these organisms, what we would call a part of their evolutionary inheritance.

But what about their genes? As far as zoologists know, there is nothing in their genome that would compel the birds to reproduce only once a year. There is no mechanism to stop them from reproducing shortly after having successfully brought up their young. The once-a-year cycle is triggered by a much simpler, genetically informed system that prompts hormonal changes resulting in an interest in reproduction only when the length of daylight passes a specific threshold. As days get longer in spring, this system triggers the cascade of behaviors that result in reproduction.

So an evolved trait (reproducing only once a year in high latitudes) depends on two distinct pieces of information. Inside the organism, there is a genetically controlled clock with a hormonal trigger (days longer than a certain time d prompt reproductive behaviors). Outside the organism, another piece of information is the fact that days of length d occur twice a year, as a consequence of the motion of the sun on the ecliptic plane. Naturally, these two pieces of information can mesh together and produce a specific behavior because of other physical facts, like the fact that reproduction takes more than x weeks from start to finish, and x is longer than the interval between two days of length d. The important point here is that you can get an evolved trait or behavior of organisms without genes that specify that trait or behavior. To the extent that stable properties of environments supply the additional information required, natural selection never had to supply it through genes.

It may seem that there is a great distance between thrushes and warblers detecting that the time is ripe for sex and complex human behaviors like building political systems and learning technology. Indeed, it is a great distance, because humans acquire vastly more information, of more diverse kinds, from their environments than other organisms, and because they acquire most of it from other humans. But the principles of information apply to the complex case as they do to the simple one. Information consists of detectable states of the external world that reduce uncertainty in the internal states of the organism. The process requires a set of possible internal states, organized in such a way that they be modified in predictable ways by the information received.[6] Genes and the complex structures that they help build obey the same principles in interacting with environments.

Rule II: Information Requires Evolved Detection

So far, so simple. But the interaction of genes with environments has some unintuitive consequences. One of them is that there is no such thing as the environment—there are only particular environments from the standpoint of organisms with particular genes. The fact that day length passes a particular threshold, at some point in spring, may have important consequences in some birds' brains. But it leaves most other organisms completely indifferent. Dung beetles carry on eating and digesting dung with the same enthusiasm, entirely oblivious to what quails and warblers find so important. And that is not because beetles are less complex organisms than birds. Often the apparently simpler animal detects what more complex ones ignore. For instance, salmon and eels can detect subtle changes in the amount of salt dissolved in the surrounding water—these are parts of the environment for those fish, and a crucial piece of information for organisms that migrate between fresh and salt waters—but such changes are not detected by supposedly complex organisms like ducks, otters, or human swimmers.[7] Similarly, closer to our everyday experience, an extraordinary variety of delicately different smells constitute the environment of a dog, yet they are all but undetectable to the more complex human brain. It is an evolved feature of salmon and eels that they detect currents and salinity, and infer the direction they should take. It is through genetic selection that some birds became sensitive to daylight duration or to Earth's magnetic field. It is also because of their specific genomes that bees and birds can detect light polarization that is blithely ignored by most mammals. Again, information from the environment affects only organisms whose genes produced the right equipment to detect that particular kind of information.

But we have great difficulty in applying this straightforward principle to human organisms. We routinely accept that humans extract all kinds of information from their environment but fail to see that this is only possible because of specialized information-detection equipment. So let me use another example. An aspect of our environment that is packed with relevant information is the direction of people's gaze. In physical terms, the relative sizes of those two fragments of the white sclera, on both sides of the iris, that are visible when we open our eyes can be used to infer a direction of gaze, which is itself used to select the object a human being is attending to. It is clear even to infants that this is an important piece of information, which commands their attention and can reveal to them what someone is paying attention to, that is, a person's invisible mental states.[8] Extracting information from the environment requires knowledge, because, strictly speaking, that information consists of cues (the two white areas of the eye) that trigger specific inferences (an estimate of the ratio of the right/left sclera area), which in turn leads, via some subtle trigonometry, to the computation of a specific direction of gaze, which itself supports a representation or a mental state, for example, "she's looking at the cat"). This rather complex computation requires not just the geometric competence but also a host of prior, very specific expectations. That is, the system cannot compute what you are looking at without assuming, among other things, that there is indeed such a continuous line between eyes and objects, that it is always a straight line, that it does not go through solid objects, that attention usually focuses on whole objects, not parts of objects, that the first object on that line probably is the one attended to, and so forth.[9] These are all pretty subtle and complicated assumptions, but they are required to do something apparently as simple as detecting where someone is looking.

And the subtlety does not stop there. Knowing where someone is looking also tells you about that person's mental states. If there are four different cookies on the table, and the child has her eyes intently fixed on one of them, which one do you think she really wants? Which one will she pick up? This kind of guessing game is trivially easy for most of us. But some autistic children have a hard time with that question, to which they give random answers. They can tell you which one the child is looking at—no problem there. They also know what it is to want something. But the link between looking at the cookie and wanting the cookie is often opaque to them.[10] It takes a special pathology to alert us to this fact—the connection between direction of gaze and intentions is a piece of information that we must add to our understanding of the scene. That the child prefers the cookie is information only if you have, again, the right kind of detection system.

What detection systems an organism possesses is of course a conse-quence of evolution. Humans constantly use gaze detection to infer each other's mental states, an immensely useful capacity in a species where indi-viduals depend on constant interaction with others for their survival. Being able to infer what other people are looking at is a great advantage when you need to coordinate your behavior with them. If you see gaze detection in this evolutionary perspective, you could also predict that domesticated ani-mals, to the extent that they interacted with humans, might be able to detect human gaze too. That is indeed the case for dogs, whose domestication in-cluded complex interaction with humans for protection and then hunting, two activities in which some minimal understanding of human intentions was an advantage. By contrast, chimpanzees can detect gaze in humans only after an excruciatingly long training, and even then their performance is not great—because their evolutionary history did not include such joint at-tention interaction with humans. That is also why domestic cats, which do not engage in cooperative tasks with humans, are generally clueless in that respect.[11]

To repeat, then, information is there only if you have the right detec-tion system—and you have the detection system because having it, or a slightly better version of it, proved advantageous to your forebears, over many generations. But here's the catch. Precisely because we have these detection systems, and they work smoothly, without us being aware of their operation, their existence is well nigh invisible to us. It seems to us, when we reflect on the way we manage to understand the world around us, that the information really is there for the picking, just waiting to be noticed, as it were.

This way of thinking, where we know about the information detected but fail to remember that it requires a detector, is called spontaneous or naive realism. It is our natural way of thinking about the way our minds acquire information, as if information were waiting for organisms to pick it up.[12] Spontaneous realism is difficult to abandon when it comes to complex social phenomena like detecting prestige or beauty or power. But even power is an abstract quality that is all but invisible to those without the right cognitive equipment. It may seem that it is not very difficult to understand who is in power in a human group—after all, some people seem to boss others around, so how could one not see that? But thinking that way is falling into the spontaneous realism trap. People's behaviors can be seen as instances of "bossing" and "obeying" only for a detection system that pays attention to behaviors as expressing preferences, that can attach specific preferences to different individuals, that can recall the relevant parts of individuals' behaviors, and that presupposes much more, for instance, the notion that rank is transitive, so that if a is superior to b and b to c, then a is above c. Young children already have some intuitions about the behaviors that make dominance manifest, long before they have much knowledge of their implications in human life.[13] So, to repeat something that does need a lot of repetition, there is no information picked up from environments without the right kind of detection equipment. Fortunately, we are now making great progress in understanding the specialized systems that help humans acquire information from their social environments, even for subtle differences in power or prestige or beauty.

Learning How to Babble, Be Good, and Menstruate

Another consequence of information detection is that the more information organisms pick up from their environments, the more complex their detection systems will need to be. Moving along a continuum of complexity, from rather simple protozoans to cockroaches to rats to humans, we find organisms that acquire more and more information from their surroundings. But the capacities of these organisms also become vastly more complex, from amoebas to rats to humans. Acquiring more information from environments requires more information in the system. In fact, it is a good rule of thumb of cognitive evolution that organisms that learn more are the ones that know more to start with. (This should be unsurprising to computer users above a certain age, who can compare systems they use nowadays to those available twenty years ago, now-vintage computers that could "learn" much less—that is, receive and process much less information, of fewer different types, from the digital environment—because they had less prior information, that is, less complicated operating systems than more recent machines.)

More complex organisms can engage in more learning than simpler ones. Learning is the general and very vague label that we use to describe a situation in which an organism acquires some external information, which modifies its internal states, which in turn modifies its subsequent searches for information. A great deal of learning is involved in most humans' behaviors. Here are a few examples of how learning unfolds in young minds.   .

BABBLING

From birth (and indeed some time before that) infants spontaneously pay special attention to speech, as opposed to other sounds, and can recognize the typical rhythm and prosody of their mother's language, perceived in a rather muffled form during the last months of gestation. In the first months of life, this leads them to pay attention only to recurrent sounds that are pertinent in their language and to ignore everything else as noise. That selective attention is reflected in babbling, which starts as a wonderfully catholic mixture of all possible sounds one can make with vocal cords, a mouth, and a tongue, and gradually restricts itself to the sounds of the local language.[14] Paying attention only to specific sounds in turn allows infants to identify the boundaries between words, a pretty difficult thing to do, as the stream of speech is generally continuous.[15] So learning takes place in steps—isolate language from the rest, isolate features of your language, isolate relevant sounds from noise, isolate words from each other. Each step, obviously, requires some previous expectations—for example, that there is such a thing as speech, that it matters more than other sounds, that there are two kinds of recurrent units, sounds, and words, and so on.[16] At each point these expectations allow the organism to orient to a special aspect of the sonic environment, and at each step these expectations are in turn modified by the kind of information that was picked up. These expectations make children attend to some properties of speech as carrying meaning but not others—they expect that the recurrent difference between ship and sheep, or between chip and cheap, may carry some difference in meaning, but they ignore the difference in the word "ship" pronounced by a man and a woman—even though the acoustic contrast is just as great. Children can acquire their native language, from interaction with other speakers, because some very specific mental systems are prepared to attend to specific properties of sounds.

BEING GOOD . . .

Children also learn about invisible things—morality is a good example. The difference between moral and immoral behaviors does not reside in any properties of the behaviors themselves. Given special circumstances, giving away your money might be criminal, and beating people may be commendable. As the moral value of actions cannot be observed, human minds have to "paint" moral qualities on behavior. How does a developing mind learn to do that in the appropriate way?

A tempting explanation would be that children observe and experience the negative rewards that accompany disapproved behaviors, in the form of punishment, and then generalize some kind of negative quality to many other behaviors. But that is not as simple as it seems. First, even infants are sensitive to antisocial behavior. They dislike puppets that clearly try to hinder or harm other puppets—and that is long before they can experience disapproval or punishment.[17] And if children actually tried to generalize from what they are told is wrong, how would they do it? Children from an early age can attend to the kinds of actions that others around them, adults in particular, seem to condone or condemn. But adult reactions are no help if you have no understanding of their underlying reasons. People tell you that it was quite wrong to attack an old lady in a dark alley and steal from her purse. Fine, but how do you conclude that it is also wrong to shortchange a blind person? One might think that this is not hard—all you have to do is to notice that, in both cases, someone used unequal strength or another advantage to exploit a vulnerable individual. But to produce that generalization, the developing child has to mobilize, at least in an implicit way, abstract notions, such as freedom versus coercion, exchange versus exploitation. To some extent, that is exactly what children do—their early intuitive understanding of cooperation and fairness helps them make sense of otherwise unpredictable moral judgments.

There is in the mind a moral learning system, a detector for morally relevant information in the environment. That much is made obvious by the fact that some people lack it. Psychopaths are those people who actually develop in the way predicted by the commonsense theory, that we could learn morality by generalizing punishment, that is, understand moral values only in terms of what rewards they bring. Individuals of that kind do realize that a range of behaviors lead to punishment, which is against their interests. They conclude from these facts that they must manage to get the benefits of whatever they want to do, while avoiding the unpleasant consequences.[18] They survive, and sometimes thrive, by exploiting others and making sure they can get away with what others intuitively find repulsive and exploitative. This peculiar syndrome has of course attracted considerable attention, and there is now a large amount of evidence concerning the specific brain activation patterns, hormonal profiles, and modes of thought associated with that behavior.[19]

So the story of the child simply picking up moral understandings from the local culture, by observation and generalization, is terribly misleading. It seems plausible only if we do not bother to fill in the blanks in that description and specify exactly what information is picked up, how, when, and by what system.

. . . AND HAVING BABIES

Now ponder the question, How do young girls learn to menstruate? It may seem a strange question, but certain aspects of reproduction do involve some form of learning. Consider the prevalence of early teen pregnancy in some parts of the United States. This was and still is clearly associated with socioeconomic status and education. Poorer young women (in the lowest quartile of income) are much more likely than richer ones to become pregnant before the age of twenty. Many social programs tried to address what was seen as a pathology, or as the result of ignorance concerning the facts of life. But they had practically no effect, and were based on dubious assumptions anyway—in modern urban environments, young women do know how sex leads to reproduction.

So, if teen pregnancy is not just an aberration, why does it occur? Large-scale studies show that many circumstances contribute to the phenomenon. One major and rather surprising factor is that young women whose biological father was (for whatever reason) absent from the household during early and middle childhood are more likely to engage in early sexual activity, and also to become pregnant at an early age.[20] The separation of parents but also the timing of separation are strongly predictive of early menarche (first period), early sexual activity, and teen pregnancy.[21] These factors remain even if one controls for the effects of socioeconomic status, ethnicity, or other social factors. But what is the connection between father absence, long before the girl's puberty, and early sexual maturation? There is no evidence that it has anything to do with a lack of parental authority (fathers laying down the law) or economic status, or of local norms, that is, young girls just imitating what is done around them. None of these factors, in any case, would explain the link between father absence and the timing of a girl's first menstruation.

A more plausible explanation, that is still partly speculative, is in terms of learning. The fact that a growing girl has no father may provide her with an indication that, in her environment, fathers generally do not invest in their offspring. If durable investment from high-value males is unlikely, and if one's own prospects are also unlikely to improve, an efficient strategy would be steeply to discount the future, increase the number of one's off-spring, and have them as early as possible.[22] That is a strategy that is mostly open to very young women, at the peak of attractiveness in the eyes of such males. These factors would converge to favor an early reproduction strat-egy, whereby a woman produces more children earlier. This explanation makes sense of many other features of the phenomenon, like the fact that young women with no fathers express more interest in infants, even unre-lated ones. This interpretation is of course not definitive, as we have to fill many gaps in the proposed causal chain—and it takes vast amounts of data to disentangle the effects of different variables. Also, it may be the case that some of the variance in such behaviors is driven by genetic differences, so that daughters tend to replicate their mother's reproductive strategies partly because they carry the same genes.[23]

Naturally, there is no need for conscious decision making here. Young women do not think in these quasi-evolutionary terms, assessing the men in the local mating market in terms of their potential costs and benefits. Rather, they respond to internal motivations and preferences, among which are sexual attraction, romantic love, a longing for children, and the satisfaction of having them. Unconscious processes attend to relevant information in the environment and favor one among the several reproductive strategies available to humans.[24]

INTUITIVE INFERENCE SYSTEMS

I mention these examples of learning to illustrate some properties of the mental systems that organize human behavior, by acquiring vast amounts of information from the environment, including of course from other individ-uals and what they do and say. Learning is made possible by a whole range of mental mechanisms that I call here intuitive inference systems (other common terms are "modules" and "domain-specific systems").[25] The "in-ference" part of the name just means that they handle information, and pro-duce modified information, according to some rules. For instance, to turn sounds into meaning, we rely, first, on a system that receives a continuous stream of speech and turns it into a largely imagined stream of discrete words with boundaries between them. Another system then identifies such abstract properties as word order, or prepositions, or case endings if you are listening to Russian, and other morphological information, and uses all this to parse the sentence, forming a new representation that specifies who did what to whom and how.

The human mind comprises a great number and a great variety of such systems, carrying out the most diverse computations, such as detecting people's line of gaze, assessing people's attractiveness, parsing sentences, telling friends from enemies, detecting the presence of pathogens, sorting animals into species and families, creating three-dimensional visual scenes, engaging in cooperative action, predicting the trajectory of solid objects, detecting social groups in our community, creating emotional bonds to one's offspring, understanding narratives, figuring out people's stable personality traits, estimating when violence is appropriate or counterproductive, thinking about absent people, learning what foods are safe, inferring dominance from social interactions—and many, many more. These constitute a rather disparate menagerie, but inference systems have some important properties in common.

First, these systems operate, for the most part, outside consciousness. We simply cannot be aware of the way we identify each word in speech, that is, retrieve it in less than a tenth of a second from a database of perhaps fifty thousand lexical items. In the same way, we do not know what exact computations take place somewhere in our minds, yielding the result that an individual is attractive or repulsive. We do not have to engage in deliberate reasoning to feel disgust at gross violations of our moral norms, like assaulting the weak and betraying one's friends. That is why we call these systems intuitive, meaning that they deliver some output, for example, the impression that a food is disgusting or that an individual is a dear friend, without us being aware of what computations led to that conclusion. All we can report is the conclusion itself—which of course we can then reason about, explicate, or justify. But the intuition did not need those reasonings.

Second, it is clear that the inference systems are specialized.[26] A mental system in young girls pays attention to the presence of a biological father, and several years later may affect the timing of puberty as well as motivations for sex and motherhood. The processes involved are probably of no help at all to figure out the boundaries between words in language. And learning how your language handles verb conjugations has presumably little effect on your moral development. These different systems are relatively isolated from each other. This is an unsurprising consequence of the principle of no information without detection. A system can detect information only if it does not detect other events. There is so signal unless you ignore noise. But what is noise for some inferences—for example, the clothes you are wearing do not change the way we understand your sentences—can be signal for another, for example, to figure out your social class or ethnicity. So each system has to focus on particular kinds of information.[27]

This is of course familiar to us from the use of computer programs, which are unlike evolved minds in many ways but share this property of being composed of different subprograms with dedicated functions. The word counter in a word processor tells us how many words are in a text. The spell-checker tells us whether they are spelled in our text as they are in a stored lexicon. But the word counter does not notice spelling errors, nor does the spell-checker tell you about the length of a text. And neither of these systems can tell you whether the words you used are common or re-cherché. These are all what we call domain-specific computations, as each system performs a limited set of operations and no others, on some particular input and nothing else.

Third, we can much better understand the way these different systems work, what they pay attention to, and what behaviors they motivate once we see them as evolved properties of our species, that is, ways of acquiring information that promoted the fitness of individuals that had them in their repertoire. This suggests that the best way to understand cognitive architecture, the different components of the mind and their relations, is to see how the components match specific problems we humans encountered in our evolutionary past. This way of connecting evolution and mental systems was the starting point of the domain of research now known as evolutionary psychology.[28] In precise terms, variants of genes that promoted a slightly more efficient, or slightly less costly, version of these little systems (through unfathomably complex cascades of gene activation, protein manufacture, gene switching, hormone release, and so forth) were slightly more likely to be replicated via reproduction. As will become clear in the chapters that follow, otherwise mysterious aspects of our mental functioning, and of its consequences for social life, can be illuminated by asking what contributions they could make to genetic fitness.

Seeing a mental inference system as an adaptation is only the starting point of a research program. This evolutionary hypothesis deserves consideration only if it allows us to predict new or nonobvious aspects of the inference system, and if we can test these predictions against observations or experimental evidence. As the systems are very diverse, so are these research programs. That is why we should not expect the new scientific convergence I describe here to yield a general theory of human societies. But it can produce something vastly more useful and plausible, a series of clear explanations for the many different properties of human minds involved in building human societies.

Rule III: Do Not Anthropomorphize Humans!

The poet and amateur naturalist Maurice Maeterlinck once described the tender emotions he could see on an ant's face as she regurgitated food on the colony's larvae, her eyes full of selfless maternal devotion.[29] He wisely stuck to his career as a poet and playwright. No student of ants, however admiring of their many qualities, would take that sort of description seriously. But it reflects a way of seeing nature that was (and is) not uncommon. Before we knew much about thunder and earthquakes, it seemed quite natural to think that some agents were behind these spectacular phenomena. But we learned to avoid this kind of explanation. The world is governed by physical laws, not by the intentions of agents. Trees grow and rivers flow, but not because they want to. As science gradually expanded our knowledge of the way the world actually works, anthropomorphism (seeing other species as human-like) and animism (seeing agents in such things as trees and rivers and thunderstorms) have been continuously receding from serious scholarship.

There is one domain, however, where this retreat from animism and anthropomorphism still meets considerable resistance, and that is human behavior. When we try to explain why people do what they do, our natural inclination is to see them as persons. That is, we assume that people's behavior is caused by their intentions, that people have access to these intentions, that they can express them. We also assume that people are units, that is, each individual has preferences, for example, for coffee over tea, so that it would be strange to ask what part of them has those preferences or how many subparts of them favor coffee. We treat people as whole and integrated persons. In other words, we anthropomorphize them.

That is just as wrong for a science of people as it was for the science of rivers and trees. Indeed, for centuries, being anthropomorphic about people has been the main obstacle to having a proper science of human behavior. The notions that people have definite reasons for behaving, that they know these reasons, that there is a control unit inside human minds that evaluates these reasons and governs behavior—all these assumptions are terribly misleading. They hinder proper research and should be abandoned.

There is of course nothing wrong in treating people as persons when we interact with them—quite the opposite. To construe others as unique agents with preferences, goals, thoughts, and desires is the basis for all moral understandings and norms. To see them as integrated, that is, with some centralized capacity for judgment that adjudicates between their possibly different goals and intentions, is also the only way to allocate blame and responsibility. It is a way of thinking that comes to us automatically and is indispensable for social interaction.

But not for science. That is, when it comes to understanding the actual causes of behavior, what we know of human minds and their neural underpinnings suggests that we should dispense with the notion of a centralized pilot, that an expressed preference for tea over coffee may involve dozens of mostly autonomous systems—in short, that we must do with minds what we routinely do with cars, look under the hood and figure out how distinct parts contribute to the general effect, so to speak. We have no difficulty understanding that this is the right approach as regards hugely complicated systems like immune function or digestion. But it is much more difficult when it comes to thinking.

The problem is, we human beings think we already know how thinking works. For instance, we assume (without necessarily making it explicit) that thinking takes place in a central processor, where different thoughts, essentially similar to the ones we experience consciously, are evaluated and combined with emotions and give rise to intentions and plans for action. All human beings have what psychologists call a spontaneous "theory of mind" or intuitive psychology, a set of systems that makes sense of the behaviors of other agents in terms of their intentions and beliefs.[30] Intuitive psychology is automatically activated when we consider behavior. We see an individual walking, then stop for a short while, then turn back and rush in the opposite direction—and cannot but infer that she suddenly *remembered* something she had previously *forgotten* and that she now *wants* to *attend to* that previous goal. The terms in italics all describe invisible, internal states of the individual, which we spontaneously imagine whenever we consider behavior. We spontaneously attribute beliefs and intentions to organisms from other species, which sometimes works, in the sense of predicting behavior, and often does not. We also do it with complex machines, especially ones that handle information, like computers.

The problem is that our intuitive psychology is not a precise and accurate description of the mechanisms of thought. Perhaps a familiar example will be of help. We routinely anthropomorphize computers. We say for instance that the computer is trying to send some material to the printer, but it does not know what type of printer it is, or it has not yet realized yet that the printer is switched on. Such statements (roughly) make sense, as they describe a situation in terms that provide some indication of what has gone wrong and what could be done about it. But if we want to understand why, how, and when such computing incidents occur, we have to use a completely different vocabulary. Now we have to talk about physical ports, logical ports, serial protocols, network addresses, and so forth. The philosopher Daniel Dennett describes this shift in modes of explanation as the transition from an intentional stance—we describe the parties at hand in terms of beliefs and intentions—to a design stance, where we talk about components and their relations.[31]

Understanding how minds work requires a similar transition from intentional to design talk, which is sometimes rather unintuitive. For instance, nothing seems simpler than the notion of belief. Some people believe ghosts exist and others do not, some people believe they put their car keys in their pockets, others believe a guitar has six strings, and so forth. But in some circumstances, this talk of beliefs can lead us astray.

Consider, for instance, the way people can act on the basis of magical beliefs that they do not actually believe, so to speak. In many experiments, psychologists like Paul Rozin and his colleagues have demonstrated that many people are susceptible to magical thinking.[32] For example, given a choice between two glasses of water bearing the labels "water" and "cyanide," they would rather drink from the former, even if they saw the experimenter pour water in both glasses from the same pitcher. There are many other experimental conditions where people have such apparently magical thoughts, for instance, refusing to don a sweater described as part of Hitler's wardrobe. Most participants in these studies are quite clear that they do not believe in magical contagion. Yet their behavior often contradicts that statement. Does it mean that they somehow believe in magic without believing that they believe in it?

We are condemned to such contorted descriptions if we stick to our common, intentional description of the mind, which specifies that there is a central belief box, as it were, where the organism's current beliefs are stored and combined to produce new inferences. Then it really seems to be the case that, despite their protestations to the contrary, people in these situations actually believe in magical contagion—a label on a glass makes the contents dangerous, a murderous dictator makes his sweaters somehow poisonous.

But there is another way of looking at all this, from a design stance. The mind is composed of many inference systems, each specialized in a narrow domain of available information. From this perspective, what happens when people see a glass labeled "poison" is that the systems that handle threat detection are activated, because the label matches one of their input conditions—a cue indicating a substance dangerous to ingest. Other pieces of conceptual information, for example, "this label is misleading," "this is all a game suggested by the experimenter," and so on, do not enter in the processing of the threat-detection module, because they simply do not match its input format. So they do not modify this particular system's inference that there is a threat in the environment. And given that one system in the mind is shouting "danger!" (or rather some neural equivalent) and most other cognitive systems have nothing to say about which glass is better (because there is no information to the effect that the other glass is in fact better), this may trigger, in many people at least some of the time, a slight preference for the glass with a reassuring label.

This all makes sense . . . but note that in this interpretation, neither the person nor indeed any part of the person can be described as believing that "there actually is poison in the glass labeled 'poison.'" This is true even of the threat-detection module, whose sole function is to make some parts of the environment salient and activate fear or defense responses, not to provide descriptions of the reasons for these responses. So we have a (somewhat) satisfactory explanation of why people prefer one glass to another. But in the meantime, we have quietly discarded a central part of our everyday psychology, the idea that behavior is explained by a person's beliefs, stored and evaluated in a central belief-management unit.

Anthropomorphism about human minds often results in the intellectual disease I shall call cognition blindness, which makes it difficult to keep in mind that the most trivial behavior requires a bewildering complexity of underlying computation. Cognition blindness used to be universal, and is still endemic, in the social sciences. To reprise one of our examples, can we describe young women who grew up in deprived conditions as "realizing" that their social environment makes it unlikely that they will meet a nurturing husband? Can we say that they "decide" to accelerate menstruation to allow precocious sexual activity? This of course would be very odd. It seems more sensible to say that some information about the social environment, like father presence, is handled by specialized systems, while other information, for example, concerning friendships or nutrition or ethnicity, is handled by other systems, and that their interaction predicts changes in preferences and behaviors.

Rule IV: Ignore the Ghosts of Theories Past

The study of human behavior is encumbered by the ghosts of dead theories and paradigms. It is extraordinarily difficult to stamp out those importunate, zombie-like pests. For instance, it seems that explaining human behavior requires that we talk about "nature" and "culture," or the various contributions of "nature" and "nurture" to our behavior. Or it may seem possible and also really important to distinguish what is "innate" from what is "acquired" in our capacities and preferences. Is the propensity to engage in warfare "cultural" or "natural"? Do the obvious differences between men's and women's behaviors result from nature or nurture? Could moral feelings be somehow natural, a product of our "biology," or are they the product of social pressure, of cultural norms?

These oppositions generally imply an antiquated vision of genetics, in which stable and inflexible genes interact with unpredictably diverse and changing environments. But that is doubly misleading. Environments do include many invariant properties, which is why natural selection can work. In fact, I mentioned one such property when I described a highly stable aspect of migrating birds' environment, the apparent motion of Earth through the seasons, which makes it possible for a genetic adaptation to limit the birds' reproduction in an adaptive manner. Conversely, gene activation can be switched on or off by other genes, by coactivators, repressors, and a whole menagerie of other nongenetic material in the gene's chemical environment. Indeed, a great achievement of molecular genetics has been to show how these multiple interactions result in the construction of highly complex traits and behaviors from relatively simple genetic material.[33]

That is why the fact that a behavior is an evolved trait of organisms, that it is a consequence of natural selection, does not mean that the trait or behavior itself is encoded in particular genes, as the example of seasonal reproduction in birds made clear. Nor does it suggest that the trait or behavior occurs invariably, regardless of external circumstances, or that it is inflexible, impossible to modify, or somehow present in organisms when they are born. It suggests only that it happens in most normally developing organisms, when they encounter conditions similar, in the relevant respects, to the environments in which their genes were selected. Very different environments would lead to very different outcomes. Migratory birds raised in a spaceship would probably reproduce at a very different pace. As we know from actual, tragic cases, children growing up in complete isolation cannot fully acquire typical human language.[34] But as long as the environments encountered include the same invariances that made genetic evolution possible, we can predict that development will result in those capacities and preferences that are typical of the species.

We can now consign the ghosts to the attic, because we have a much better understanding of how minds learn from environments. Here I described some typically human behaviors, like learning the language spoken in one's group, or adapting one's sexual behavior to the social environment, or inferring people's intentions from their gaze, as probable consequences of human evolution. As we go through many other typical human behaviors, particularly those that contribute to building human societies, we shall see how in each domain it makes little sense to try to mention nature and nurture, as if those terms had a stable meaning.[35] It makes even less sense to talk about human "culture" as a real thing in the world.

The Positive Program

A proper social science should answer, or at least address, the pressing questions I listed at the beginning of this chapter—to sum up, why do humans engage in those social behaviors, like forming families, building tribes and nations, and creating gender roles? The best way to answer such questions is to do science, because the best way to understand the world in general is to do science. Humans never invented anything that goes as deep as scientific investigation into understanding why the world is the way it is, nor have we found any other way of seeking knowledge that gets it so consistently right. Doing science is also difficult and frustrating, and in many ways goes against the grain of our spontaneous ways of thinking.[36]

There are of course many skeptics who think that human societies and cultures just cannot be studied the scientific way. Some see the social world as just too complex to be successfully explained in terms of simple and general principles. Others, in a more radical way, state that human meanings or beliefs belong to a special domain of social or cultural things, which is forever closed to scientific explanation. I shall not dwell too much on these debates, because the best way to counter these conceptions is simply to demonstrate that there are indeed scientific explanations for particular social phenomena. Then, the philosophy will follow the science, as it usually does.

The following chapters chart some elements of this naturalistic science of human societies, from the way we form groups to the way we interact in families, from human attraction to religious notions to their motivation to create ethnic identity and rivalry, from the intuitive understanding of economics to their disposition for cooperation and friendship. This should not imply that we now know all there is to know about those topics—far from it. But we can already perceive how they make more sense in the context of human evolution. There is great promise in that vision, some would have said even grandeur, if we can make progress in explaining human behavior as a natural process.

Six Problems in Search of
a New Science

What Is the Root of Group Conflict?

Why "Tribalism" Is Not an Urge but a Computation

OBSERVERS FROM OUTSIDE OUR SPECIES would certainly be struck by two facts about humans. They are extraordinarily good at forming groups, and they are just as good at fighting other groups. There is no species in which organisms can do so much through collective action. There are few species where so much collective effort is aimed at attacking other groups and defending the group from such attacks. No human population is immune from potential ethnic rivalry and conflict. These can escalate into full-blown civil war and genocide. It should suffice to mention racial antagonism in the United States, the history of pogroms in Europe, the murderous conflict following the dissolution of Yugoslavia, and the innumerable ethnic wars in Africa and their culmination in the Rwanda racial massacres to provide some idea of the scope and intensity of such conflicts.

We have names for the phenomenon, like "nationalism" and "tribalism," suggesting a strong urge in human beings to side with their village, their clan, their nation, against the other side, strangers or foreigners. But saying that humans are strongly tribal does not explain anything. This is where seeing human behaviors from another species' viewpoint, or from an evolutionary standpoint, can be of help, as this perspective raises "why?" questions, such as, Why are individuals committed to their group? Why do they persist in that commitment when it might be to their advantage to defect from their group? How can groups survive at all, as cohesive units, in the face of individual, divergent interests? Why are groups often locked in intractable conflicts even when all parties realize there is little profit to be expected from prolonged rivalry? Why do group conflicts, especially ethnic ones, often flare up in outbursts of extraordinary violence? How can that occur between groups that had coexisted in peace for decades or centuries? From an evolutionary perspective, having very high group solidarity and intergroup conflicts is just like having claws on your feet or antlers on top of your head—something that requires an explanation in terms of what it did for organisms over evolutionary time.

Invented Nations?

The idea of a nation implies that each state corresponds to a community of people united by traditions, cultural values, language, and the idea of a common past. This is obviously a very modern idea on an evolutionary scale. There have been modern humans for more than a hundred thousand years, but states are a recent invention, a few millennia at the most. But if we try to understand groups and group conflict, it makes sense to start with nations, because they highlight how humans find certain kinds of group identity both self-evident and compelling.

Many new nations appeared on the map of Europe in the nineteenth century, including Germany and Italy as unified polities, but also dissident fragments of previous empires, like Hungary and Serbia, as well as new-comers like Estonia. That was the age of the Romantic ideal of nations as polities based on a common culture and language, themselves the consequences of common descent. The idea was that states should correspond to those "natural" and "ancestral" communities—rather than empires put together by conquest, modern nations would be based on the natural affinity and solidarity of people with shared ancestry and traditions. Elite Romantic movements had emphasized supposedly specific cultural features found among the common folk and had described modern nations as the unfolding of these cultural traits. From this perspective, sometimes called "primordialist," Serbia and Lithuania and Italy were already there, so to speak, as potential nations. What they had lacked, beforehand, was the political opportunity to constitute themselves as states.[1]

Against this picture, some "modernist" historians and anthropologists argued that the nations were in many cases constructed by the states. That is to say, once you have a state you start noticing or emphasizing, or in some cases deliberately creating, some common features in the populations that live under that state. From that perspective, the anthropologist Ernest Gellner, for instance, described the emergence of nationalism in largely functional terms, as the outcome of modern industrial society, arguing that modern, bureaucratic states require a class of low- and mid-level clerks with administrative skills, as well as a common language for administration, and some plausible claims to legitimacy. In Gellner's view, nation-states satisfy all of those needs. State-sponsored schooling trains the bureaucrats. The unification of a language out of disparate idioms (as happened for instance in Germany and Italy) supports communication. The state is all the more legitimate if it is seen to be founded on common cultural values and to include populations of common descent.[2] Myths of origin bolster the feeling of common destiny, anchoring the groups in a more or less fantasized past, a Golden Age to which the ethnic group could return once it regained sovereignty as a nation.[3]

This functional account suggested that most Romantic claims to ethnic authenticity were largely instrumental, that in fact many were made of whole cloth. That is to say, if the political goal was to unify a particular region and turn it into an efficient polity, one could always find a convenient myth of origin or some similarities between dialects to turn that region into an ethnicity with a common language, and therefore into a nation crying out to be born into political existence. For instance, some historians argue, there was no unified Norwegian language before the elites created it, and few people would have identified themselves as Estonian before their elites managed to carve out an independent Estonia. In the same vein, historians had great fun puncturing the "invented traditions" of some European nations, showing, for instance, that Scottish tartan and British royal rituals, commonly described as archaic and authentic, had been invented during the nineteenth century by people who assumed that any decent nation should possess relics of its past customs.[4]

This description of "constructed" nations, however, was much exaggerated—mostly because of its focus on a limited place and time, the European empires in the nineteenth century. In other places, and long before the emergence of modern bureaucratic states, people had seen an intuitive link between language, ethnicity, and polity. Despite the complexity of conflicts between regional states over millennia, Chinese people assumed that their empire should include all peoples of Han culture, and the Koreans and the Japanese thought the same way. In places as different as the Greek city-states and the Yoruba kingdoms, people had a notion of ethnic identity that was largely based on language and traditions, long before nationalism in the modern sense.[5]

This raises the question, Why are these commonalities so important? Why do they matter to people? Indeed, even if the "modernist" picture had been right, even if nations had actually been built by elites from disparate communities, we should ask, Why did people find those identities compelling? Why were they motivated to defend this (allegedly spurious) ethnic heritage? Why would the elites' machinations actually convince the populace? The reasons why all this (to some extent) worked, why people found ethnicity convincing, cannot in fact be found in models of ethnicity. The answer lies in a much more general phenomenon, to do with the construction of collective action and stable groups.

Ethnification as Recruitment

Nations are often based on ethnicity, but ethnicity itself is a mystery, or it should be. Ethnicity is the notion that a certain group of people share common interests and should unite toward the realization of common goals, by virtue of shared traditions, often language, and in most cases descent. We should not think of ethnicity as a sign of political immaturity, as a primitive phenomenon characteristic of political order before large nations, democratic institutions, and modern communications. Ethnic conflict can reemerge in formerly unitary republics, populist nationalistic politicians often work their way to prominence through democratic channels, and mass communication has made xenophobia much easier to transmit. Far from being a transient phase in human history, ethnic strife seems to be a baseline to which social groups often revert.[6] The mystery here is (or should be) that so many people, around the world, find this notion of an ethnic group natural and compelling.

When considering, say, the violent dislocation of Yugoslavia or the atrocities of Rwanda, we tend to see them as a combination of very specific historical accidents with long-lasting suspicion or grievances between groups. The historical accident is, in many cases, the disappearance or weakening of the legitimate state.[7] That was the case, for instance, in most of central Africa in the 1980s, in Yugoslavia in the 1990s, in Somalia more recently. People in Rwanda or the Balkans had for many generations been nurturing deep grievances and a hatred of neighboring groups that was only ready to burst once constraints on people's expression were relaxed. In the case of the Balkans, it would seem, authoritarian regimes (of the Ottoman and Austro-Hungarian empires) followed by totalitarian socialism were only temporary blocks on the slope leading to open confrontation.

This description is suggestive but also misleading, because descriptions of ethnic conflict often assume what ought to be explained, namely, that people already see themselves as members of groups with common goals and interests, and that they feel the motivation to support their own group against rivals. So ethnic strife occurs between collections of individuals that share interests and goals, know that they share those interests and goals, and are ready to commit themselves to some collective action in pursuance of these objectives. But social processes are not that simple, as conflicts between European groups illustrate.

For instance, there are and were clear ethnic categories in the Balkans, a place where people have identified themselves, for a long time, as, for example, Croat, Serb, Romanian, and so on. But that does not mean that such identities always and everywhere denote groups.[8] Specialists of ethnic conflicts like Rogers Brubaker emphasize this distinction between ethnic categories and ethnic groups. People routinely use categories, the world over, as a way of partitioning the social world into different classes of people—you are a Serb and I am a Croat, these people are Londoners or Glaswegians, and so on. The existence of categories does not always mean that people in these different categories form groups, that is, a collection of people that act in concert toward common goals. In most circumstances, often for a very long time, people can maintain ethnic categories without ethnic groups.[9]

In some specific historical contexts, actual groups do coalesce—for example, when the Serbs think that the Croats are threatening and must be contained (or vice versa). And the emergence of groups out of categories is precisely what we should explain. In specific circumstances, people who belonged to different categories but lived in smooth coexistence, and could peacefully interact everyday, become staunch enemies and may engage in extremely violent behavior. As many outsiders comment in cases of ethnic conflicts, this development often comes as a surprise, even to many of the participants, who rightly saw their situation so far as an example of what Brubaker calls "ethnicity without groups."[10] A standard explanation for this change is to assume that the hostility was dormant, that people secretly harbored hostile feelings toward other groups, until someone or some event broached the ancient quarrel, as between the Capulets and Montagues. But this is all entirely ad hoc, and it ignores (or takes as self-evident) the very mechanism we should explain, that of recruitment for collective action.

As Brubaker points out, ethnicity is not a fact, it is a process that turns social categories, momentarily, into cohesive groups. And it is a cognitive process, whereby a mass of external information is interpreted in ethnicized terms, and the costs and benefits of participation are tweaked in a way that previous attitudes did not always predict.[11] How and why this happens should therefore be understood in terms of cognitive capacities and motivations. I think it can be explained only by abandoning the narrow domain of ethnicity for a while and considering the processes of group formation in much more general, evolutionary terms.

Are Humans "Groupish"?

Humans, as we all know, are strongly motivated to form and join social groups—that much is uncontroversial. That group living itself is beneficial, for some species, is not an evolutionary mystery. But what we need to explain is what particular skills and motivations were selected as a way of getting individuals to act efficiently in groups. Difficulties come up when we want to understand what the underlying psychology is, what explains "groupishness," as Matt Ridley called it.[12] Over the past fifty years, a large social psychology literature has documented many aspects of this "in-group bias" in modern societies. It is not limited to just preferring members of one's group but pervades many domains of cognition. For instance, people do not recall information about out-groups and in-groups in the same way. They are much more distressed by disagreements with in-groups than with out-groups. They empathize with in-groups more than out-groups, especially in the context of interaction with out-groups.[13] People are not convinced by statements uttered with a foreign accent—indeed, even infants seem to distrust potential playmates with an unfamiliar accent.[14] A host of studies document the physiological effects of interaction with out-groups, from cardiovascular to hormonal processes and stress reactions.[15]

Humans are so attached to forming groups that they seem to create group solidarity, and conflicts between groups, on the flimsiest of excuses. A salient aspect of groupishness is the contrast between the often tenuous link between members of the group, their actual connections, and strong motivations to defend the group and attack rival ones. History records many examples of this, like the famous Nika riots of 532 CE, in which supporters of rival chariot racing teams, the Blues and the Greens, attacked each other and then destroyed about half the city of Constantinople.[16] European football supporters and sports aficionados the world over provide examples of this form of tribalism.[17]

From all this, social psychologists inferred that humans were, indeed, so spontaneously groupish or tribal that they would favor their group even if the group was entirely arbitrary, and even if groups were arbitrarily constructed by an experimenter. A spectacular demonstration of the phenomenon was Henri Tajfel's "minimal group" paradigm, where people were assigned to two distinct groups, A and B, or blue and red, or any other meaningless label, on the basis of clearly arbitrary criteria. People grouped together had nothing particular in common; in fact they did not interact during the experiment. They were just told that they had been assigned to group A or B, and which group each other participant belonged to. After a while, in an ostensibly unrelated task, they were asked to allocate various goods and tokens among all participants. The result, replicated many times, was that people invariably tended to favor members of their own group. The effect remains the same, whatever the value of the goods, the familiarity of the task, or the cultural background of the participants. The phenomenon even extends to largely unconscious processes, as people without realizing it tend to sidle in the direction of in-groups rather than out-groups.[18]

These results seem to demonstrate a strong, automatic motivation to benefit one's group, however spurious the group. That is precisely the point, for psychologists, of making the groups minimal. Members of such groups have nothing in common except the label they have just been given. One can even randomize group assignment right in front of the participants and still get the effect. That is to say, one seems to favor members of one's own group in situations where there is no possible reason for doing so except that these individuals have just been described as belonging to the same category as oneself, which seems irrational.[19]

But do the results really show indiscriminate groupishness? As social psychologists after Tajfel pointed out, so-called minimal groups are not actually that minimal. In the experimental paradigms, participants allocate goods or symbolic good points to all others, and (this is crucial) they expect to receive similar goods or symbolic good points from all others. So, in the context of the study, their own welfare or self-esteem depends on the fact that they will be favored by others.[20] This sheds a different light on the apparent irrationality of the effect. It is not that people wrongly think of an arbitrary grouping as a real social group. Rather, as the psychologist Toshio Yamagishi pointed out, the mistake is for participants to assume that they are engaging in a social exchange interaction, in which people can reciprocate favors. Participants, knowing that they will allocate goods to others and receive goods allocated by these same others, intuitively (and wrongly, in this case) infer that they will receive more if they give more. As this reciprocation heuristic is constantly activated in real in-group situations, people spontaneously apply it to whatever in-group situation they experience. Empirical evidence confirms this. When participants allocate goods to others but do not receive goods from these same others, the in-group bias disappears.[21]

So groupishness is not a blunt instinct to follow the herd, so to speak. People behave in ways that seem to favor in-groups because they implicitly use a social exchange heuristic, a set of assumptions about how the social interaction that is presented to them (evaluating different individuals or allocating resources between them) is a form of reciprocal cooperation.[22] Obviously, they need not do that consciously. All they experience is the value they attribute to particular individuals. But the computations are taking place, away from conscious access. Which is why it makes sense to explore in more detail this hidden world of mental computation that makes groups possible.

Coalitional Psychology

Before understanding social groups, we have to understand how humans form alliances and how alliances recruit members. Social groups, from small cliques of friends to entire nations, and from tribes to trade unions, exist only because individuals are motivated to join and remain. As human beings, we find it quite natural that humans belong to groups, and we may even be tempted to think that groups existed before individuals, so to speak, that they have an existence of their own. But if we step back and take an evolutionary perspective, the construction of groups is not at all a straightforward process, as it requires that different organisms with different genomes manage to overcome conflicts of interest and set up mutually advantageous social interaction.[23]

So it makes sense to start with alliances, with the fact that several individuals behave in ways that enhance each other's welfare. Coalitions are found in many social species, especially in apes. But they are small-scale, often unstable, and limited in their scope. By contrast, human alliances can include large numbers of agents, can persist for generations, and are in fact ubiquitous, extending to all domains of human behavior. Coalitional processes may be found at many different levels of organization, such as political parties, street gangs, office cliques, academic cabals, bands of close friends. They can include thousands or millions of individuals in ethnic or national coalitions. Informal, small-scale networks of solidarity are also found in organizations that should not, in principle, depend on them, like large corporations that have formal rules for coordinating the behaviors of many agents, all of whom occupy well-defined positions with explicitly understood chains of command and information transmission. But in most corporations there are cliques of employees who volunteer information and help to each other.

The influence of coalitions is also salient in armies. Most armies take as their fundamental unit of action a small group of ten to twenty-five men. Such units are usually grouped in larger groupings of two to five hundred agents. Experience shows that soldiers are most efficient when they operate in the context of a small group of people they know well, a group within which there is maximal solidarity and very high personal trust. The larger units command some loyalty and usually come with a sense of shared identity, but they are less efficient for achieving specific goals.[24]

Finally, political parties are of course the prime example of large-scale coalitional affiliation. Party members set great store by loyalty. In most legislatures, party-based groups generally impose votes along party lines. A politician who switches parties is generally seen not just as someone who changed his mind or preferences but as a renegade by his previous group and not much of a recruit, usually, by his new one. This notion that change of opinion or allegiance is by itself morally repugnant is so intuitive that we rarely wonder why it is so.

There is no social history or ethnography of any human community that does not mention people joining forces for common goals, creating and maintaining rival alliances, and punishing defection. This is so pervasive in human social interaction that the point seems banal, which is precisely why the contribution of cognitive science is crucial here. Three-dimensional binocular vision, too, seems straightforward until you try to describe how it is done by neurons. In the same way, it takes cognitive science to understand that alliance building is problematic and requires specific skills.[25]

So where does coalition building come from? Naturally, the kinds of alliances and coalitions that humans build are enormously diverse. But that diversity is, precisely, made possible by an underlying set of psychological capacities and preferences that seem to be part of our evolved cognitive equipment—the set of specialized capacities and dispositions that constitute what evolutionary psychologists, after Leda Cosmides and John Tooby, now call a coalitional psychology.[26]

For their survival and reproduction, humans have always needed extensive support from kin, but also from nonkin. Such support is essential in the most diverse domains of social interaction. Throughout human evolution, fitness depended on the extent to which nongenetically related individuals could be recruited for help. Social support is one of the crucial mechanisms that provide fitness, because so many human activities involve intense cooperation—think of hunting, trade, defense against enemy groups, and probably shared parenting. There is also some evidence that prehistoric humans helped the less fortunate members of their groups, as individuals seem to have survived illness and injury—which suggests that they were provided for, even though they could not contribute to hunting or group defense. Modern conditions, of course, provide many more contexts in which humans can exert their dispositions for recruiting support from others.

For alliances to emerge, all participants must hold specific mental representations. First, they must represent a certain goal, such that it is better obtained through joint effort than individually. Second, each individual must expect that the other members have a roughly similar representation of the goal. Otherwise members of the alliance would not expect coordinated effort from others. Third, one must discount one's own costs in working for that common effort. Coalitional work, like all collective action, requires behavior that may seem altruistic, that is, conferring a benefit to others at one's own cost. But the cost is offset by expected future gains from the collective venture—whether that expectation is warranted or not. Fourth, one must expect that others, too, will discount their effort. Fifth, one should expect others to have that expectation about oneself. Finally, one should represent all costs (or benefits) by the rival coalitions as benefits (or costs) to oneself, so that one is motivated to increase (or decrease) them.[27]

This may seem unduly technical, but experimental evidence suggests that human minds easily and unconsciously perform all these coalitional computations. For instance, people readily interpret benefits to the coalition as benefits to the self.[28] Also, memory experiments demonstrate that people presented with conversations between unknown third parties automatically attend to who is allied to whom, even though they are not instructed to do so—and often do not realize they did so. This information is then retrieved from memory more easily than other features of the interaction.[29] That is, people may have fuzzy memories of exactly what was said, but they recall quite clearly who opposed whom. So it seems that our coalitional psychology includes an alliance-detection system, that is, a system that spontaneously attends to information about the social world that suggests a specific pattern of solidarity or affiliation between some individuals.[30]

People also monitor commitment and defection, because investing resources and effort in a coalition is disastrous if others free ride on the common achievements, or if they defect when it is their turn to invest, for example, if members of your platoon run away when the going gets rough. That is why we are often so eager to detect signals of commitment in others, such as public statements that one is a member of the group, or actual contributions in time, effort, resources. Indeed, the worry about defection is such that people spontaneously interpret the mere possibility for others to opt out of the common enterprise as a form of betrayal worthy of moral condemnation.[31] That also explains why people attend to the difference in status between longtime members of a coalition and newcomers. Specifically, the fact that someone is a newcomer triggers the intuition that she may be free riding, as she receives the benefits associated with membership but has not had enough time to contribute much to the common cause. For instance, from the first day you join the U.S. Marines, you benefit from the prestige attached to that military corps. But you have not yet demonstrated your willingness to put yourself in harm's way for the defense of the group. We know from experimental studies that this combination of getting benefits but not paying costs triggers an automatic process dedicated to the detection of free riders. That may explain why old-timers in many groups are often so aggressive toward the greenhorns, and submit them to severe hazing or initiations.[32]

These computations occur in cognitive systems that deliver definite intuitions (for example, "these people are a group and they are against us") and motivations (for example, "we should accommodate them/attack them," "we need allies," and the like), while the underlying computation is inaccessible to conscious inspection. Equally unconscious are the motivational effects of these computations, the fact that they trigger specific hormonal release and engage emotional neural systems.

Are alliances always against other alliances? Human coalitions are often competitive, pursuing their goals against those of other coalitions. A great part of coalitional psychology consists in mobilizing support against others. Why should that be the case? It is not necessarily because human nature is intrinsically antagonistic. No, the competitive nature of coalitions lies in the fact that they constitute attempts to recruit social support, and support is what economists call a rival good. The more someone gets, the less is available for others. Any interaction that promises to deliver social support for some individuals will lead the others to form their own network, lest they remain without partners.

Coalitional Intuitions Feed Stereotypes

So we cannot attribute to humans a brute instinct of groupishness. Motivations for joining and defending groups seem to entail a more sophisticated psychology, a set of systems that help us recruit social support from unrelated individuals. But, if this is valid, we may have to discard some entrenched views about social groups and conflicts between groups. In particular, we may have to rethink the common assumption that people are motivated to help their own community, and conversely to work against another, rival group, because they share some values, ideas, goals with other members of that group.

Consider ethnic-racial prejudice, stereotypes, and discrimination. It may seem that one discriminates against people from another ethnicity because of hostility against them. You exclude Irishmen from your workforce because you dislike them. Hostility in turn seems to be based on negative representations of a social category—you dislike Irishmen because you see them as loud and violent drunkards. In the vocabulary of social psychology, we would say that stereotypes (a representation of what is common to members of a social category) lead to attitudes (an emotional reaction to interaction with those people), which in turn lead to discrimination (behaviors that actually diminish those people's welfare).

The picture is intuitive enough that many people consider, and some social scientists argued, that discrimination could decrease if attitudes evolved, and attitudes could evolve if more people could realize how unfounded stereotypes are. Once you get to know actual Irish people, your experience gets in the way of stereotypes. That was the basis of the "social contact" idea of social psychologists, that race relations in the United States, for instance, would get better only if more whites were in personal interaction with blacks.[33] But is that the case? The evidence is mixed—in fact contradictory. For instance, the U.S. Army is a highly integrated organization, with people of different ethnicities in constant interaction from the lowest to the highest echelons. It is also the place where people report the highest satisfaction with interracial relations in the nation.[34] But that cannot be the effect of mere contact. After all, slaves and their owners in the antebellum South were in constant, personal interaction. We know of many other situations of ethnic diversity where contact with other groups is intense, and so are suspicion and even hatred toward the other group.

So there may be something wrong in the assumption that stereotypes lead to attitudes that lead to behaviors. In evolutionary or functional terms, the assumption seems a tad mysterious. What is the function of these stereotypes in the first place? In other words, what advantage is there to think of members of other groups as essentially lazy, stupid, or dangerous? One common view, called Social Identity Theory, assumed that people somehow needed to think of themselves and their groups as better than others. But the notion of such a need was completely ad hoc, postulated just because it would explain ethnocentrism—in fact explaining it by saying that people needed to be ethnocentric.[35]

To see how stereotypes and behaviors are related, consider James Sidanius's work on ethnic discrimination in different countries. Studies show persistent attitudes that associate the stereotypical information about some people as lazy or incompetent or violent with emotions of fear and contempt. There are also observable discriminatory practices. The question is which of these facts causes the others. The standard answer was that the stereotypes were the source of all this particular evil. By contrast, Sidanius and his colleagues described this as a dominance situation, showing that ethnic groups and "races" are perceived from the start in terms of competition for resources.[36]

So, is there discrimination because of negative stereotypes, or are stereotypes a way to justify hostility toward a competitive alliance? A test of these distinct explanations lies in patterns of discrimination. If we assume, for example, that racism toward blacks in the United States is motivated by stereotypes and identities, we should expect all members of the target group to be equally discriminated against. By contrast, in the coalitional dominance model presented by Sidanius, males would be the prime targets for prejudice, as they constitute a more salient threat to one's coalitional advantages. Males from the minority group would be seen as most dangerous, since they are more likely than females to engage in violent revenge. This also suggests that men from oppressive groups would engage more readily than women in discriminative practices. This is what Sidanius called the "subordinate male target hypothesis."[37] The evidence supports that hypothesis. Experimental studies and observations show that faces of black men are much more likely than female faces to activate stereotypes (incompetence, violence), attitudes (rejection), and specific emotional responses (fear). The process is automatic. It does not depend on explicitly accessible thoughts about the group. For instance, many American subjects are much faster at identifying a gun as a gun, and slower at seeing a tool as a tool, after seeing a briefly flashed black man's face, but not after seeing a woman's. Reverse effects obtain with white faces. Also, actual discrimination seems to target minority men more than women, for instance, in rental prices or car insurance premiums.[38]

This would suggest that racial categories, in the United States, are construed by most people as a proxy for rival coalitions. That is in fact confirmed by a series of striking experiments by Rob Kurzban and colleagues. The motivation for these studies was that social psychology experiments had shown that the "race" of displayed persons was automatically encoded in psychology experiments. No matter what explicit instructions are given, no matter how irrelevant race is, no matter how much extra cognitive work has to be done, American participants always seem to recall the racial identity of the faces they see during an experiment. Why would that be the case? One possibility is that race is just perceptually salient, a feature that is just out there, so to speak, available for minds to attend to. But perceptual differences, like other pieces of information from the environment, are there only for systems attuned to detecting them. Would that mean that humans are equipped with a system that detects racial differences? From an evolutionary standpoint, that would be very odd. It is only very recently (on the scale of genetic evolution) that humans have been in contact with different-looking others. In the conditions in which human minds evolved, you only encountered people with phenotypes highly similar to you own. So a race-detection system is unlikely to have evolved—in contrast to our spontaneous attention to age or sex.

One way to demonstrate that people unconsciously encode race as a substitute for coalition is to create experimental materials in which they have to attend to a coalitional rivalry, for example, between team A and team B, where both teams include members of both races. If race encoding is really perceptual, participants would have no difficulty recalling the race of all the agents presented. If, by contrast, race is coalitional, they would make errors and mistake whites for blacks, or vice versa, when these different-race individuals were members of the same team—and that is precisely what happened in these studies and in subsequent replications.[39]

So it would seem that in many cases our standard view has things backwards. That is, what drives people's behavior is a coalitional situation, where it seems advantageous to try and keep members of other groups in a lower-status position, with distinctly worse outcomes, on the basis of an intuition that the welfare of different groups is a zero-sum game—we can only gain if others lose. Far from being a case of stereotypes leading to strife, it would seem that rivalry between groups is intuitive, immediately obvious to many people, while negative representations of members of the other group are a highly relevant way of explaining those intuitions. In this model, stereotypes do not cause behavior but provide a relevant interpretation of that behavior, for those engaged in discrimination or other coalitional behaviors.[40]

Building Large Groups by Signaling

Accepting that our coalitional psychology helps us build groups, that is, collections of individuals with common goals, in potential competition with other alliances, it remains to understand how this can work with very large groups. That is a problem, because the coalitional mechanisms I have described here seem to apply very well to small groups where members personally know each other and can gauge each other's contribution. But coalitional dynamics go much further than that, extending to large groups that number thousands of agents. That is possible because people can signal coalitional affiliation. Dress, accent, gestures can be used to make manifest that we belong to a specific coalition.

We can better understand these codes and symbols if we replace them in the context of the models of signaling, developed over forty years by biologists and game theorists.[41] A great variety of animals emit signals, that is, they convey information to other organisms about their own state or dispositions. Here one must distinguish signs or indices—cues from which one can infer something about the environment and other organisms—from signals that were designed to provide information.[42] A deer's smell is an index of the animal's presence, but when a male deer rubs his forehead against a tree, that is a signal addressed to other males. A female chimpanzee's genital swelling is not just an effect of her being in estrus, it is a signal designed to attract potential mates. There is of course no other designer here than natural selection, as these signals contribute to fitness.

We know that ethnicity can be detected in myriad ways, as people from different places or groups dress or talk or cook differently. But some of these indices, some of the time, are used as signals. They convey that the person is a member of that particular social category, but they are used because of that effect and are addressed to other members of the group or to outsiders or to both. When black Americans adopted Afro hairstyles in the 1970s, or when some Muslims started to don traditional Middle Eastern garb in Western Europe in the 1980s, they were signaling ethnic and cultural affiliation. In many tribal groups, specific ornaments, tattoos, scarifications, or other body modifications signal membership of the group. In some modern contexts, people would say that adoption of ethnic markers is motivated by pride in their particular cultural heritage or affiliation. But this raises the question why that pride should be expressed in that particular way at that particular juncture, and that is best explained as a form of signaling.

Signaling has some consequences, predicted by formal game-theory models and confirmed by observation. For one thing, some signals are reliable, as they convey true information about the sender, in circumstances where knowing the truth is advantageous for the receiver. For instance, it is in the interests of female chimpanzees to attract males when they are fertile rather than at other points in the menstrual cycle, and it is in the interest of the males to reserve their mating efforts for fertile females. By contrast, when two organisms' interests diverge, signals can become deceptive. A cat will arch its body and raise the hair on its back to appear larger than it actually is to an enemy. Dishonest signaling of this kind can lead to an arms race between deception and detection, as senders become better at conveying false information while receivers become better at seeing through the subterfuge. Another dynamic emerges when the fact that some organisms send signals forces others to do the same. For instance, toads may signal their vitality and size to potential mates by croaking in low-pitched tones that reliably indicate a large body. But if one toad does it, all have to do it—silence becomes an index of low quality. Not signaling means that you have nothing great to signal.[43]

It sometimes happens that by signaling their affiliation to a particular alliance people make it impossible for other coalitions to recruit them—and that is often the point. For instance, Philip of Orleans, who changed his name to Philip Equality during the French Revolution, voted in favor of the death penalty for his cousin Louis XVI. This signal of loyalty to the new regime was all the more significant, as it made it impossible for the man to ever return to, let alone lead, the core royalist camp.[44] The point of such signals is not just to demonstrate affiliation to one alliance but also to burn one's bridges with the others. If that is a function of coalitional signaling, we should expect highly stable coalitions to favor commitment signals that are irreversible. Indeed, ethnic identification in tribal contexts often consists in tattoos, scarifications, and other forms of body modification that leave permanent traces. This may also explain why tattoos were mostly favored by criminals, until recent fashions diluted their significance. As detailed in Diego Gambetta's *Codes of the Underworld,* criminals often need to signal to potential partners that they cannot leave the underworld and return to lawful economic activities, that they are committed to illegality because they have nowhere else to go, so to speak. As long as the larger social world looked askance at highly visible tattoos, the individual who bore them signaled to his associates that he was unlikely to defect.[45] The same goes for gang tattoos, which make it impossible for the individual ever to be recruited by another gang. Such computations need not be conscious—people in general, and criminals in particular, do not usually bother to think in terms of game theory and biological signaling. But the individuals concerned intuitively understand that getting scarified or tattooed conveys much stronger commitment than knowing the tribal songs.

Each signaling behavior has an effect not just on receivers but also on other potential signalers, as the economist Timur Kuran pointed out.[46] Consider the situation of a man living in a Muslim country, where most religious officials and devout men wear a beard. The individual may choose to adopt that style, too. By doing that, he may be signaling adherence to the religious faction, presumably as opposed to other people's more secular attitude. But his behavior is also changing the signaling landscape for all other men. That is, he has increased, ever so slightly, the relative ratio of bearded to shaven men, which may be one of the ways in which other individuals estimate the relative costs of demonstrating religious affiliation or not. As a result some of these men may themselves change behaviors, and so forth. At each step, the perceived costs of shaving or not shaving are changing, which of course affects the probability that other individuals will make one of these two choices. This process of "reputational cascades" leads to bandwagon effects, as many people very quickly seem to adopt a new style or express solidarity with a particular cause. When this occurs, it seems to outsiders that many people have changed their preferences and convictions in lockstep. But the interpretation is misleading. We do not need to assume a change in most people's minds when they react to the changed costs of signaling—as each person who adopts the signals increases the reputational cost to all others of not doing the same.

Computations of Violence

Ethnic hostility often explodes in violent confrontation, from the European pogroms against Jews to outbursts of Shi'ite versus Sunni hostilities in the Arab world, to the many civil wars of Africa to the series of riots and massacres that accompanied the partition of India. Violence attests to passion, but passion should not be interpreted as inchoate emotion. On the contrary, outbreaks of ethnic violence show that rage and aggression are the result of complex computations inside minds.

A symptom or consequence of these computations is that ethnic violence, although it may seem very diverse, often takes a predictable form. At first sight, there seem to be few common traits among episodes of ethnic aggression. Some violent events occur in the context of civil or national war, others in peacetime; some involve only small posses of determined aggressors, while in other cases the violence engulfs an entire region or country. But Don Horowitz, in his survey of ethnic riots throughout history, shows that there are also many common features, besides the obvious ethnic nature of the confrontation. That is, when ethnic violence breaks out, it generally follows a pattern that is remarkably similar across very different places and times.[47] To start with the obvious, ethnic riots occur in places where ethnicity is a clear marker of social identity, where most people belong to one of several mutually exclusive groups that claim common descent and shared interests, and where most people know where the others live. Equally obvious, ethnic strife requires that most people consider the welfare of their city or nation as a zero-sum game, where the prosperity of one group entails diminished resources for the others. Finally, people generally know and transmit memories of salient historical events and grievances that confirm this zero-sum perspective and the other groups' evil designs. In many of these places, social interaction between members of the different groups is relatively easygoing, at least not overtly hostile, which makes the outbursts of violence all the more baffling to observers.

But the most surprising finding is that, in many cases, ethnic riots follow a highly similar "script." They start with an apparently minor episode, such as a scuffle between youths, an angry reaction to some sporting event, a dispute between landlord and tenants, and in general a conflict that is circumscribed and might indeed be construed as entirely nonethnic. Such events often have no consequences at all. But in some cases they are amplified by rumors of more deliberate acts of aggression, or of preparations for such aggression, to the effect, for example, that "they" attacked and killed children, or they poisoned the wells, or they are planning to kill or evict "us." After these rumors start circulating among the group, there is a period of cautious and restricted interactions with the others, a period that is unusually and ominously quiet. After a few days, another very small incident occurs, escalates into a proper fight, people start recruiting others for help, entire neighborhoods mobilize and storm the stores and houses of the "others," people may try to kill the fleeing victims and start chasing members of the enemy group wherever they are known to live. This is when the worst violence occurs, as people do not just beat members of the other group but may also try to shoot them, burn them, bury them alive, mutilate and torture them. The victims, including women, invalids, and infants, can hardly escape their fate, and their pleas for mercy are ignored and mocked.[48]

Ethnic violence is not an uncontrolled outburst of rage. The fact that it takes such predictable forms means that some common processes are shaping these violent interactions, and that participants have psychological capacities and preferences that make it possible for them to engage in these acts in a coordinated manner.

This seemingly counterintuitive conclusion, that violence occurs precisely because of complex computations, is also clear in the often gruesome tactics of insurgents during civil wars. From antiquity down to the civil wars of twenty-first-century Africa, observers have remarked that there is more violence and cruelty in civil war than in organized national conflict.[49] While armies mostly engage in limited violence and proceed in a fairly predictable manner, agents of civil war and insurrection seem to pursue violence in unpredictable ways, often, it would seem, for the sake of inflicting harm rather than achieving rational military objectives. Bands of insurgents, or militias on the side of the government, loot and sack villages, murder or rape or mutilate individuals suspected of siding with the other side. Assassinations and assaults in the context of ethnic struggle often take on grotesque forms, as in the famous cases of "knee-capping" in Northern Ireland.[50] During the Rwanda racial war of 1994, the attempt to exterminate the Tutsi population led to previously unknown levels of cruelty, with the systematic killing, torture, or mutilation of vast numbers of civilians. What explains the intensity of violence?

One factor is that gruesome violence is a form of signaling. That is, perpetrators know that crimes are more likely to be reported, and to instill the desired level of terror, if they are public, salient, massive, and emotionally compelling. This explains not just the level of violence but also some of its bizarre details. For instance, executions during the Rwanda massacres often took on specific, arbitrary forms directly inspired by traditional methods of animal sacrifice. The perpetrators' intuition would be that they needed to use methods that, in the local cultural context, would constitute the most striking and efficient signals.[51]

Another important factor is that the victims of violence, in both ethnic rioting and civil wars, are seen as dangerous aggressors. Just as Jews were described as cockroaches by the Nazis, the propaganda in Rwanda described the Tutsi as insects, dangerous parasites whose presence threatened the survival of Hutu communities. In the same way, the rumors that precede ethnic riots often describe "us" as facing a terrible and imminent threat. "They" might poison us all, kill all the children, burn down our houses. All observers concur that the main emotion is fear, indeed terror, at the prospect of what might happen soon. For third parties, this is paradoxical, as the group targeted is generally a minority. But threat detection is a powerful motivator. The tragic fact that previously peaceful individuals can engage in atrocities is very often rooted in fear.[52] This may recruit specialized capacities that we know exist in human minds, notably the mental systems engaged in detecting predators and attacking prey. Humans evolved as successful hunters because of a sophisticated understanding of predator-prey interaction, which in particular motivates violent aggression in situations of threat.[53]

An important additional factor is the intrinsic uncertainty of combat in situations of insurgency or civil war. Typically, the actors engaged in a civil war include the incumbents (government forces and their allied militias), insurgents (in more or less organized associations of combatants), and a mass of noncombatant civilians. Each combatant side has very little information about the other's actual fighting power, and, crucially, neither side can be certain about the local population's support. This creates a situation in which people are more likely to engage in indiscriminate violence, affecting noncombatants, women, children, and the rest, as a way of signaling the strength of their side and the risk of joining the other.[54] Villages that harbored some combatants will be razed or burned down, as a signal of the cost of collaboration. Uncertainty about allegiance motivates the combatants to implicate the civilians as much as possible, and if possible to make it impossible for them to join the other camp. During the Rwanda massacres, militia warriors would often force civilians to participate in atrocities, for instance, to murder their neighbors or friends, on pain of seeing their own children or relatives killed.[55] This is compounded by the fact that, in many civil wars, the breakdown in law enforcement makes it possible for many people to eliminate enemies or exert revenge under the cover of political activity. The political scientist Stathis Kalyvas summarized these various aspects of civil war in a model that shows how the occurrence and level of violence depend on a few variables, such as information about the enemy, the signaling needs of different agents, the cycles of revenge, the control structure within the group, and, most important, the security dilemma faced by either insurgents or incumbents, the fear that the enemy will certainly attack us if we do not strike first.[56]

The Shadow of Primitive Warfare

Humans routinely attack each other, either individually or in groups, in all observed societies. Does that mean that there is a universal "aggressive instinct" at work? A large popular literature describes humans in terms of such instincts. This conveys the idea that violence occurs when people let off some form of accumulated aggressive energy that was present for some time and needed some outlet—and the outlet may not be chosen with great care. These hydraulic metaphors are also involved when people consider that young men, for instance, need wars or violent exercise to release their natural aggressive instincts. The metaphors are difficult to resist, because they are entrenched in many common ways of thinking about anger and violence. We also say that people "exploded" or that they released "bottled up" feelings in an "outburst" of anger. That is also the way people describe anger in many of the world's languages, as energy that comes in variable amounts seems to be getting stronger as we ruminate on what made us angry in the first place, and seems much easier to manage after expressing it through words or actions, as if some pressure had indeed been relaxed inside some system.[57] But these are, obviously, just metaphors. Mental processes do not consist in hydraulics, in systems accumulating high pressure and releasing it through safety valves or violent explosions. There are no such things in human minds.

If you consider what happens in minds in functional terms, as solutions to recurrent problems faced by organisms, the notion of an aggressive urge makes little sense, especially not for highly social organisms like humans. Among the many, many different kinds of social interaction that humans experience, some are such that violence is a valid strategy (for example, if a rival shepherd tries to steal your sheep, if a stranger snatches your infant from your arms). In some situations, violent aggression is a losing strategy (for example, if a large group of invaders tries to steal your cattle). In other circumstances, mutually beneficial cooperation is the best option (for example, if several groups of pastoralists need access to a single source of fresh water). If humans were moved by a context-free, irrepressible instinct for violence, they would not create any difference between these situations and would attack any time they could. By contrast, individuals that could attack but also refrain and cooperate, depending on the situation at hand, would fare much better, and therefore leave more offspring with similar dispositions, than simple-minded aggressors. In other words, whatever genes make humans (like other organisms) conditional aggressors have a better chance of being replicated than genes that result in undifferentiated aggressiveness.[58]

Was the war between groups really a constant feature of ancestral life, and as such a likely source of selection pressures that would have molded our cognition and motivations? Hobbes described humankind as perpetually mired in the war of all against all, before a sovereign could impose peace with subservience as the prize for safety.[59] Rousseau described peaceful primitive communities that only acquired aggressive motivations, as well as all sorts of other ills, as a consequence of property and civilization.[60]

For a long time cultural anthropologists tended to side with Rousseau's perspective, as they witnessed very little crime and tribal warfare in the small-scale societies they studied. People would not risk their lives to acquire at great risk what a little work could produce so easily. Also, many social scientists at the time thought that the very low population densities of early foraging groups in prehistory would have made warfare extremely unlikely, in contrast to the crowded conditions brought on by agriculture.[61] But a more careful look at the evidence, and more systematic studies, put paid to this depiction of peaceful ancestral communities. Anthropologists had observed little crime, but in proportion to the size of the communities, the impact of homicide was actually very large. One is more in danger of being murdered in some tribal societies than in the least safe of American inner cities.[62] Also contributing to an underestimation of warfare was the fact that tribal conflict had been significantly dampened by colonial authorities by the time anthropologists could engage in fieldwork.[63]

As regards relations between bands or tribes in ancestral conditions, several factors suggest that conflict must have been an ever-present possibility throughout our evolutionary history. One reason is that humans are a territorial species. In all known human groups, people assume that a given polity goes with a given territory, that there is a clear demarcation between space that is "ours" and space that is "theirs." True, early humans were nomadic foragers. But nomadism generally consists in movement within a well-defined and delimited space. Indeed, nomadism requires extensive territories, which leads people to resent the passage or intrusion of other groups.

That is why there is little truth in either the Hobbesian or the Rousseauist vision of early human societies—or rather, both are true of different aspects of early societies. Our tribal past certainly included both intense cooperation (within small groups) and trade and peace (between groups), as well as frequent aggression for murder, looting, and abduction (both within and between groups). The error of both visions was to think of humans as driven by unconditional, inflexible instincts toward war or peace. What makes humans go to war or cooperate is not stable, general, and context-free preferences for aggression or for peace, as Hobbesians and Rousseauists believed, but a set of conditional mechanisms that weigh the value of either strategy, given the current environment.[64]

What kind of warfare did result from such conflicts? The anthropological and archaeological records suggest that it almost certainly did not take the form of pitched battles, with large armies marching against each other in an open field. Such organized confrontations occasionally occur in some tribal societies, but they are mostly for show—the two parties size each other up, exchange threats and insults, and often return home without exchanging blows. The more serious confrontations take the form of sudden raids. These do not require a large number of combatants—indeed, they exclude mobilizing large groups, to avoid detection. The plan is to kill as many of the male members of the other group as possible, take whatever plunder is available, and in many cases carry off women or slaves back to camp before the enemy can summon up sufficient resistance.[65]

This form of intergroup aggression was observed in virtually all the small-scale tribal societies observed by anthropologists before state powers made warfare more difficult. It was also ubiquitous in forager groups such as Northwest Coast Native Americans and even Australian Aborigines, who made very few tools but many weapons. Eskimo foragers, too, managed to make weapons and armor with scant resources. Pueblo villages were built on hilltops so that aggressive maneuvers could be detected. Oral traditions but also archeological evidence confirm that this state of potential attack was the case for as long as evidence is available, from rock paintings in Australia or North America to the mass graves of concussion victims in Paleolithic Upper Egypt. Even though foraging bands and tribal societies knew about peacemaking and the value of alliances, they were also beset by the constant threat of tribal warfare. This was not a marginal aspect of their existence, as demographic studies in some groups suggest that 5 to 20 percent of males may have died in combat.[66]

From all this, we can construct a plausible picture of primitive warfare. Primitive here does not mean archaic. The term simply denotes the fact that this form of conflict does not require vast killing technology, or the mobilization and coordinated action of vast numbers of soldiers, like modern wars. In this sense, ethnic riots, urban guerrilla battles, many insurgencies, and most certainly murderous conflicts between street gangs are examples of primitive warfare. These are all highly similar forms and require the same mental capacities.

First, such primitive warfare is based on the advantages of asymmetry. In some cases, this is simply an asymmetry of information, as members of one group know when they will attack, but the victims have to fear attack at any point. Also, in raids of this kind, people only engage in combat when they think they have a clear numerical advantage, in which case they display extreme violence in the treatment of enemies. In this respect human raids against neighboring groups are not very different from conflicts between bands of chimpanzees. When the aggressors realize the field is not in their favor, they just abandon the fight and run away, for fear of being massacred.[67]

Second, because of this asymmetry, many raids are successful. Even if the attackers do not manage to inflict severe losses on their target, at least they can often withdraw without suffering too much damage. As a consequence, it is a sensible strategy to try and prevent raids by preemptive strikes. Since the members of a group that is becoming numerous and powerful may at any point attack us, we should attack them before they get too powerful. But since they too know that, it is even more likely that they will attack us first, which makes our preemptive strike even more justified, and so forth. On top of that, it makes sense to launch a preemptive strike that will not just discourage the potential enemies but in fact incapacitate them— hence the desire to kill rather than just overpower members of the other group. The perceived need for preemptive strikes to avoid preemptive strikes, which is called the security dilemma, would seem to guarantee a perpetual state of tribal warfare or at least intergroup suspicion. Obviously,

in many cases people managed to escape the security dilemma by contract-
ing alliances between groups, often formalized by gift exchange and marriage. But these are only temporary guarantees against the potential threat
of preemptive strikes.[68]

Third, primitive warfare was and is overwhelmingly a male operation.
Men organize raids and carry them out against the violent opposition of
other men, and a frequent outcome of these raids, if not their main goal, is to
abduct or assault women. This asymmetry extends to physical violence in
general, not just in the context of group rivalry. Physical violence is potentially extremely costly for women, because injuries may drastically reduce
their reproductive potential, while it is much less so for men. Conversely,
the capacity to protect one's own is crucial to a man's reproductive value.[69]

Still, why would anyone participate in group conflict, given the danger? Fighting would seem to undermine the imperative of maximizing fitness, so that a motivation for bellicose enthusiasm would be much less
successful in genetic reproduction than a capacity for sneaky desertion. But
this may be misguided, in terms of both costs and benefits. In terms of costs,
raids are less dangerous for the aggressor than pitched battles. The benefits,
which in many cases of tribal warfare included abduction and slavery, may
outweigh the probability of serious injury or death. That is, a modest likelihood of death may be offset, on average, by a large gain in prestige, power,
and access to reproduction. That is why, as John Tooby and Leda Cosmides pointed out, there is a tacit "social contract" in warfare operations.
Even if a cost is certain, the identity of which individual will bear the cost,
that is, who may die or be seriously injured, should remain unpredictable.
On the other hand, a share in the benefits, in the event of success, should be
certain, otherwise people will often refuse to participate.[70]

The long past of primitive warfare makes sense of features of present-
day conflicts that are otherwise puzzling. Humans are extremely good at
handling the various cost-benefit calculations that make primitive wars possible. For instance, neither street-gang members nor ethnic rioters need to
study social science to have a clear understanding of the advantages of inflicting extreme violence on noncombatants. They intuitively understand
the power of gruesome signals and the desirability of preemptive strikes.

They guess that males and females cannot be recruited in the same way or threatened in the same manner. Civil wars and ethnic riots take on predictable, and tragic, forms because they occur between organisms with highly similar capacities for group aggression.

Detecting Diversity

Most modern societies are thankfully spared the horrors of ethnic rioting and civil war. People of different ethnic or racial categories coexist, and in many contexts these categories do not coalesce into groups, in the sense of having internal organization and collective goals. But diversity does not make ethnicity, ethnic signals, or a sense of affiliation vanish altogether. Indeed, in many countries modern migrations do not lead to integration or assimilation into the host population but on the contrary usher in a renewed salience of communal identity and ethnic demarcation. As a result, urban multiethnic conditions create a novel situation for human minds, one in which individuals can be easily identified as belonging to distinct and exclusive ethnic categories, but they do not reside in distinct territories, and the extent to which they are committed to their group is uncertain.

What is likely to happen in such conditions? We might be tempted to think, and some traditional social science would have argued, that it all depends on stereotypes about different categories, based on adherence to particular political ideologies. It would seem obvious that stereotypes, for example, that a particular category of people are aggressive or lazy, would govern people's interaction with members of that category. There is of course some evidence for such effects, which in the past led many social psychologists to conclude that interaction with others was, indeed, largely a matter of top-down information processing, that is, of prior expectations and concepts. But actual interaction is more complex, and it suggests that human minds are better-designed information processors than this traditional picture would suggest. For one thing, we know from many cases (cognition about race is one clear example) that people's intuitions, for instance, that members of a certain group are coalitional rivals, often trigger stereotypes rather than derive from them.[71] Also, in cases where people have both a stereotype about an individual as a member of a category and specific information about that individual, the latter generally overrides the former. In other words, stereotypes, for example, about the loud drunk Irishman, do not in fact stop people from noticing the actual distribution of loudmouths and drunkards, and sober discreet individuals, in their social environment. Stereotypes are often discarded when they stand in the way of efficiently navigating the social world.[72]

So we should not assume that what happens in diverse modern environments depends solely on large-scale politics and ideologies. In fact, as the studies on stereotypes and individual information suggest, interaction between groups may well be determined by more humble factors, by the way people's cognitive systems acquire information from direct encounters with the social world.

Consider this. In modern diverse environments, one comes across a large number of individuals (apparently of distinct ethnicities) with whom one can engage in diverse social interaction, from sharing space on a sidewalk or on a bus, to engaging in anonymous economic exchange, to collaborating in a workplace or being associated in some collective action. How do these multiple encounters, these microepisodes of social interaction, impact our mental representations of categories and groups?

The impact of repeated encounters between individuals is not much studied, mostly because it is very difficult to do so. The effect of multiple encounters on a mind may be highly specific to a person's conditions, which will make sampling more difficult. To compound the difficulty, these effects are largely unconscious, so that simply asking people explicit questions about their perceptions of the social environment will not be enough. Yet, we should focus on those microprocesses, because they are crucial to understanding the dynamics of ethnic diversity in modern environments. This is of course a (mostly) speculative claim, because there are very few such studies so far. But there are reasons to think that representations of groups are indeed influenced by the quality and frequency of encounters with ethnic others, what could be called the ecology of modern ethnic diversity.

People's alliance-detection systems probably pick up available information, in their social environment, about the presence of different kinds of people that may belong to distinct coalitions. Some of the information consists in indices like accent and phenotype, which reveal ancestry or community. It also includes signals, for instance, ethnic markers like dress, body ornaments, religious symbols, and so forth. People in an urban environment are likely to pick up information about the relative size of ethnic groups, the number of people they seem to include. This is likely, because we know that in other domains human minds automatically, and most of the time unconsciously, produce statistical representations of their environment. Foragers estimate the frequency with which different places afford different kinds of food, on the basis of dozens or hundreds of visits to each place. Shoppers expect different stores to be more or less expensive for different categories of food. Drivers expect parking places to be more or less abundant depending on the neighborhood and time of day. These expectations are the result of automatic statistical inferences from multiple episodes. They are examples of natural sampling, a form of intuitive statistics that comes easily and intuitively to human minds.[73] One may speculate that the same applies to ethnic identities. Each and every day, many people in modern societies probably compute similar, and similarly unconscious, statistics about members of different ethnic categories.

It is also likely that people can pick up information about the cohesiveness of their own coalition and its potential rivals, the extent to which people in each group are committed to each other's welfare. Cohesiveness is required for coalitional success, and the more cohesive a rival coalition is, the worse that is for us. That is why people in coalitional conflict tend to emphasize similarity ("we are all the same") as a way to convey solidarity ("we all share the same goals") and therefore coalitional cohesion ("we shall all fight on the same side"). In contexts of warfare, people wear identical uniforms or highly similar, group-specific tattoos, scarifications, or makeup, which also convey this message. And as I mentioned earlier, people tend to adopt ethnic markers, such as dress that is diagnostic of one ethnic group, all the more readily when there is ethnic rivalry. It would be surprising if our coalitional psychology did not register that kind of information. For instance, over the past thirty years, Muslims in European countries have become more noticeable than they used to be, as many men and women adopt traditional dress that is uniquely Muslim, probably contributing to the perception of that particular ethnic-religious category as numerous and cohesive.[74]

There is some evidence that such intuitive statistics about coalitional membership have some effects on people's attitudes. For instance, Robert Putnam has argued that increased diversity often correlates with a decrease in social trust, in the extent to which one considers most others as trustworthy.[75] In the United States, the effect can be measured at the level of states and counties, which is not very precise. But a study from Denmark shows the same effect, with much more precise measurements, by using data about the actual numbers of individuals from distinct ethnicities that reside at different distances from each participant in the study. Here one's confidence that others are generally trustworthy decreases as a function of the actual number of foreigners that may be encountered, suggesting a clear effect of social ecology.[76]

Our automatic coalitional statistics may even impact our health. Many studies have shown that people's health and subjective well-being are affected by ethnic relations. Members of minority groups in general have poorer health outcomes than the host population.[77] But that difference holds even when one controls for the effects of the obvious factors, such as socioeconomic status and access to physicians, suggesting that dominance relations have their own effect on people's health.[78] How could that be the case? One possible pathway is stress. From multiple physiological studies, we know that encounters with members of other ethnic-racial categories, even in the relatively safe environment of laboratories, trigger stress responses.[79] In everyday life, minority individuals have many encounters with majority individuals, each of which may trigger such responses. However minimal these effects, their frequency may result in accumulative stress, which would account for part of the health disadvantage of minority individuals.[80] This possible explanation is supported by another, apparently paradoxical observation, that minority individuals are often in better health if they reside in nonintegrated neighborhoods, in ethnic enclaves, than if they are mixed with the host population.[81] This so-called ethnic density effect is not well understood. One possible explanation, in terms of our intuitive computations of coalitions, is that minority individuals in minority neighborhoods simply experience fewer encounters with other-group individuals, and therefore suffer less accumulative stress.

Although partly speculative, this interpretation of modern ethnic diversity is congruent with what we know about the psychology of intergroup relations in general. From tribal rivalry to modern nationalism, from peaceful claims of identity to murderous ethnic rioting, we can make sense of very diverse, occasionally paradoxical behaviors in terms of evolved capacities for coalition building and coalitional defense. Humans depend on group cohesion and continuity for their own individual welfare and survival. The stakes are very high, which explains why the evolved systems trigger very powerful motivations, and why the outcomes of these unconscious computations take the form of pride, suspicion, rage, or hatred.

What Is Information For?

Sound Minds, Odd Beliefs, and the Madness of Crowds

KILLING BABIES, DRINKING BLOOD, AND occasionally eating
fetuses is what middle-class English people do, or at least what some of
them were described as doing during the 1980s—when not submitting their
own children to bizarre forms of sexual abuse in the context of Satanic rituals. What had started as a rumor became a public crisis, when more and
more cases were reported and some children volunteered their own testi-
monies of horrendous rituals. Local authorities were flooded with anonymous accusations. Some social workers managed to persuade the authorities
that the children should be taken away from their homes. After more careful
police investigations, it turned out that there was no evidence for any of
those alleged episodes of abuse, Satanic or otherwise.[1]

Penis-snatching outbreaks have afflicted African and Asian countries
for at least thirty years. In many places, people report that dangerous individuals have the power to steal a man's genitals by simply looking him in the
eyes, shaking hands, or pronouncing special magical words. It is also said that
crowded places like markets and bus stations are these dangerous people's
favorite hunting grounds. What usually happens is that someone in these
public places suddenly shouts that his penis got stolen, pointing to a specific
individual as the perpetrator. The suspect is quickly surrounded by an outraged mob. This can lead to a summary execution. With more luck, the supposed thief is led to the local police station. Although people are certain that
a penis was stolen, they are equally confident that searching the accused
would be futile, as he probably disposed of the evidence through magic.[2]

The world over, as far back as historical documents can be found, human groups have experienced episodes of panic of this kind. Witchcraft accusations are another example. An individual claims to have suffered magical attacks on the part of some relatives or acquaintances, and enlists the support of the community to make the alleged witches confess their wrongdoing.[3] The craze can give rise to a large and lasting social upheaval, as in the witch hunts of Elizabethan England, where hundreds of suspected witches were tried and found guilty (not necessarily in that order).[4] Why do these panics occur, and why do people hold such strange beliefs to start with?

These episodes are salient illustrations of something much broader, an entire domain of culture in which people's passions are triggered by information of extremely low value (to be clear, there simply are no witches or penis thieves). To rephrase T. S. Eliot, it would seem that humankind can bear a lot of unreality. The emergence and success of such beliefs is puzzling, given the fact that our minds were shaped by natural selection as efficient learning machines. There seems to be a failure of engineering, if human minds are so susceptible to information of such low value. That is why so many anthropologists in the past wondered, Why would people believe these things? But this raises another, often neglected question, Why does all this matter to people? Why are people motivated to tell others about such events? Why would they participate in the witch hunt? Why the madness of crowds?

Mysteries of Junk Culture

The age of information began some time between five hundred thousand and one hundred thousand years ago, when humans started exchanging information at a rate unseen in any other species. Many kinds of organisms send and receive signals, within or across species, but humans do so orders of magnitude more intensively. The natural environment of human beings, like the sea for dolphins or the ice for polar bears, is information provided by others, without which they could not forage, hunt, choose mates, or build tools. Without communication, no survival for humans.[5]

Which makes it all the more surprising that a lot of the information transmitted is of no use whatsoever, and that people can become so passionate about that useless information. I refer to this vast domain of apparently useless information as "junk culture." One may find the term a tad negative, but the inspiration here comes from the "junk DNA" of molecular biology, those large segments of our genetic code that seem to convey no useful information. As it happens, geneticists are now finding that some parts of so-called junk DNA have specific and important functions.[6] In a similar way, perhaps junk culture can in fact be explained in functional terms.

Examples of low-value information are not hard to find. The anthropological record is replete with odd theories that seemed to combine a fierce grip on people's imaginations and motivations with a complete lack of useful content. One could of course fill volumes with illustrations of what Kant would have soberly described as the exercise of reason beyond the confines of experience.[7] They are also catalogued in such classics as Mackay's *Extraordinary Popular Delusions and The Madness of Crowds*, and many subsequent compilations.[8] But it may be more useful to describe more precisely the boundaries of junk culture.

Classical anthropologists did consider the mystery of junk culture, which was variously called symbolism, magic, or superstition, and often described as typical of "other," that is, non-Western cultures, whose members perhaps conducted their cogitations in a way that was radically different from "ours." Theirs, it was claimed, was a primitive mentality that would be closer to free association than to causal reasoning. This would, for instance, explain beliefs in magic. Some believe that eating walnuts cures brain illness, because a shelled walnut resembles an exposed brain. Or burning a lock of someone's hair would make that person sick, as the hair was a part of the person. In this view, relations of similarity or contiguity mattered more to the primitive mind than relations of cause to effect.[9] But this form of cognitive relativism soon lost much of its grip on the anthropological imagination, as it turned out that beliefs in magic (and other apparently unfounded beliefs) were common in Western societies. Conversely, familiarity with faraway places showed that people in those cultures managed their everyday affairs with as much common sense as your seasoned Western philosopher. Scholars had first thought that peasants were cognitively different; after peasants moved to the cities, Africans were described as magical thinkers; then, as Africans visited Europe, it was in the jungles of Papua New Guinea or Amazonia that anthropologists imagined there must be a radically different way of thinking. The progress of mass transportation slowly killed relativism.

Granting that most people in all cultures occasionally indulge in what used to be called primitive thinking, it remains to explain why they would do so. To do this, we must go back to the point emphasized in the Introduction, that humans are learning machines and require a lot of sophisticated learning devices to detect useful information in the environment.

Good Design: Learning Requires Knowledge

It takes some reflection to appreciate the gargantuan amounts of information required by any moderately complex behavior, let alone by human interaction with conspecifics and the environment. A great deal of that information, in the case of humans, develops through communication with other humans and experience of the social and the natural world.

Decades of experimental studies of infants and young children have confirmed that human minds require a lot of prior knowledge to acquire so much information. It is also clear, from those developmental findings, that the prior knowledge takes the form of expectations about very specific domains of information. For instance, infants who acquire their first words expect them to denote whole objects rather than parts or colors. They expect solid objects simply to remain separate rather than merge into each other after a collision. They also expect the number of objects in a bag to remain the same, if no one adds or subtracts items from the bag. These principles appear long before the child can actually manipulate objects. At a later stage, young children expect the insides of animals, more than their external appearance, to be what makes them behave in a particular way, so that a cat made to look like a dog is really still a cat. More subtly, infants expect an agent to behave on the basis of what she believes, not just on the basis of what is true. Even twelve-month-old infants assume that people may have false beliefs. All these (and many, many more) show the operation of expectations about solid objects, numbers, animals, and minds.[10] There are also expectations about social support, about social exchange and cheating, about fairness and morality, about hierarchy, about friendship, about predators and prey, about facial expressions, about contagion and contamination—the list could go on.

Domain-specific expectations enable efficient information gathering. For example, young children intuitively expect animals to move by themselves, and they also expect that internal states, like intentions and beliefs, explain those movements. This allows the developing mind to focus on specific information, for example, what is in front of the animal, such that the animal would try to approach it, and ignore equally perceptible but irrelevant facts, for example, what lies downstream as the explanation of why the river is flowing. As philosophers and cognitive scientists have noted, a cognitive system without specialized expectations would be forever mired in an astronomical number of irrelevant contingencies.

Even when adults explicitly convey new information about some object, children filter that information, depending on their assumptions about the adult's mental states. Gergely Csibra and György Gergely found that infants are sensitive to an adult's pedagogical intentions, expressed, for instance, by a request for joint attention. In such conditions, infants expect the information conveyed to be about a whole category, not just a particular instance.[11]

To sum up, then, even from the earliest stages of cognitive development human minds seem designed to acquire useful knowledge about their environment. I must insist on the word "useful." We should not assume that human minds are designed to acquire true information about their natural and social environments. That is an important difference. Just because something is a fact does not mean that humans are equipped to find out about it. Conversely, many of our intuitive expectations lead us to false beliefs. For example, we humans tend to see living species in essentialist terms. We assume that there is some internal quality, found in all members of the species, that explains how it is distinct from others. This implies that there is an unbridgeable gap between any two species. A giraffe is a giraffe, and a horse is a horse, and never the twain shall meet. But that happens to be false. Giraffes are horses, in the sense that the two species are in fact linked by a continuous line of reproduction, if you go far enough into the evolutionary past. An essentialist belief in strictly separate species makes natural selection difficult to comprehend, and by comparison makes creationist ideas very intuitive.[12] But essentialism about species is our spontaneous assumption, and it may be part of our evolved cognitive equipment, as it provides a convenient way of organizing information about different animals and predicting their behavior, which is of great adaptive value, even if the main hypotheses are misguided. So it is important to remember that the human mind is not always philosophically correct or scientifically accurate. The assumptions it contains may not be true, but they are useful.[13]

Usefulness, then, refers to selective pressure. We have expectations about gaze as an index of mental states, because we are organisms that need to understand other agents' mental states to survive. We have different sets of intuitive principles for man-made objects and natural beings, because we are toolmakers and must understand the connection between the shape of objects and their functions. We have social expectations because we need social support. As we shall see, we have moral intuitions because we depend on fair exchange to prosper. In each case, having these cognitive dispositions made our ancestors more successful than others at reproduction, which is precisely why they turned out to be our ancestors.

This makes our original question even more pressing. If humans are designed to acquire useful information about their environments, why, then, do they produce and absorb junk culture? One explanation might be that we are, precisely, designed to acquire most of our information from other people. Humans acquire only some of their immense store of knowledge about the world from direct experience. We have the evolved disposition to seek information from others, and in many domains to use that information as the basis for our own decisions. So, is it possible that this disposition goes too far, so to speak, and makes us vulnerable to low-quality information from others?

Good Design: People Are Not Gullible

One of the strongest beliefs held by human beings, in most cultures, is that humans are gullible . . . especially other people, of course. To most of us, this seems the most natural explanation for the obvious fact that other people's beliefs are often shockingly misguided. But is this the case? Leaving aside the self-flattery, are humans really gullible?

For a long time, cognitive and social psychologists assumed that there was a general disposition in human beings to be unduly receptive to information from others, especially persuasive others. And many studies seemed to demonstrate the power of suggestion and persuasion. For instance, in the 1950s Solomon Asch carried out experiments that were widely reported as demonstrating the power of suggestion. Participants had to provide an answer to a perceptual question, for example, which of several lines was the longest, the correct answer to which was very clear. Before they could express this opinion, however, other people in the room, confederates of the experimenter, would voice the wrong answer. A striking result was that some participants would agree with them, apparently persuaded that they should not believe their own lying eyes.[14] In a similar way, memory researchers in the 1980s showed how one could persuade people to accept as true a childhood event that had never happened, for example, that they had been lost in a shopping mall. Using doctored photographs or enlisting the help of complicit siblings, one could even persuade people that they did recall the event in question.[15]

Although such effects became part of the received wisdom of psychology, the evidence was rather more complex than these summaries would suggest. When the cognitive scientist Hugo Mercier systematically reexamined evidence for human gullibility, he found that the oral tradition (and many textbooks) had very much distorted the original findings.[16] For instance, in Asch's famous conformity experiments, most people actually did not revise their perceptual judgments, and even Asch himself commented on that result. Indeed, his aim was to study what one would have to do to overcome this ingrained obstinacy. In the same way, most subjects in the fake childhood memory experiments did not recall the invented incident, and many did not even agree that it could have happened. Again, the point of the original studies had been lost. What memory researchers were trying to show was that, given a large amount of fabricated "evidence" and the help of reliable third parties like relatives, people might be convinced of the reality of such events. That was an important and useful point, against the then-widespread craze for "repressed memories," in which self-styled therapists would use constant suggestion and even hypnosis, for months on end, until their patient recalled some nonexistent episode of abuse.[17] So the studies showed that memory suggestion could work, but (and this is the crucial point) only if it was particularly intense. They did not show that people's memory was easy to fool—quite the opposite.

The same goes for many of the alleged effects of persuasion. It may seem that people are easily persuaded by flimsy evidence and weak arguments, and many studies support that impression. But the effect occurs only in highly constrained conditions, for instance, when the participants have no other evidence than what is presented to them, when they have reason to trust the experimenter, and, most important, when the information in question is of no concern to them. When the information matters, or when people have several sources to choose from, people are much more difficult to persuade.[18]

We should not be too surprised that human minds are rather reluctant to absorb information from others. The idea of humans as belief-acceptance machines that will accept most information conveyed by others was always a tad mysterious, if you considered it from an evolutionary angle, as gullibility would result in exploitation. Humans have vastly more sophisticated communication capacities than any other species. People communicate to change other people's mental states, for instance, to create a belief in someone's head ("Look! There's a crocodile!") or create a motivation ("Could you pass the salt?"), and many other subtler processes. Now, to the extent that one can modify another organism's beliefs, it is of course possible to change them to one's advantage. Signaling is likely to turn deceptive, as the organisms' interests diverge. Male domestic pigeons puff up their neck feathers to mislead females about their size and vitality.

This would result in a massively devalued form of communication, a situation in which no one can benefit from verbal communication, because it is overwhelmingly unreliable. But that is not really the case, for a simple evolutionary reason. Deception may be adaptive, if you can exploit others, but then it becomes adaptive for others to develop the symmetrical weapon, the ability to see through deception. There is an equilibrium when capacities for deception and detection are roughly equivalent. But that equilibrium is unstable. Any organism that is slightly better than others at deception will gain an advantage, so that it will transmit its deceptive skills to its offspring, until these skills become the population average. But then an increase in detection skills becomes adaptive, and in a similar way will gradually become the average. This kind of arms race between deception and detection is common in nature.

In the case of human communication, the arms race consists in a competition between the capacity to make one's utterances persuasive, on the one hand, and the ability to protect one's beliefs from deception, on the other. Dan Sperber and colleagues called this latter capacity "epistemic vigilance," the motivation and capacity to detect and discard unreliable information, and to check arguments for their validity.[19]

The need for epistemic vigilance explains many aspects of human communication. For one thing, people are attentive to the sources of information and maintain an estimate of a source's reliability, which affects how they process information. Conversely, the more suspicious the information, the better we recall its source as unreliable. Also, people automatically pay attention to discrepancies or contradictions in other people's statements, with consequences for the status of information as well as the source, which are both tainted, as it were, and barred from serving as guides for behavior. The same goes for causal gaps or non sequiturs—anyone who tells us that kangaroos jump high because the Australian weather is hot is inviting such suspicion.

Some rudiments of epistemic vigilance appear early in cognitive development. Infants, for instance, seem to be sensitive to the difference between expert and novice agents. Later, toddlers use cues of competence to judge different individuals' utterances, and mistrust those who have been wrong in previous instances, or those who seem determined to exploit others, or more simply agents talking about something they cannot possibly know, such as objects they cannot perceive.[20]

Most important, experiments show that people are quite good at judging (and accepting) valid arguments. Confronted with different kinds of information, especially in matters that they care about, people are generally very good at selecting relevant evidence and at choosing valid arguments rather than incoherent ones. Indeed, Mercier and Sperber argue that the emergence of reason, an evolved capacity to consider arguments in the abstract, results from the adaptive advantage we gained from being able to extract the best knowledge available in what others can provide, as well as to discard dubious advice and incoherent information. In other words, the adaptive function of reasoning is not solitary consideration of the world. It is a social tool, a set of capacities we need to convince others, to bring them around to our preferences and choices, and conversely to detect and explain what is valid and what is not in other people's imperfect arguments.[21]

All this shows the paradox of junk culture in even starker relief. Psychologists have gathered large amounts of evidence for a series of cognitive systems geared to acquiring useful, that is, fitness-relevant information about the world, especially from conspecifics, and ensuring that the information is of sufficient quality. This seems to be a straightforward consequence of cognitive evolution. In the same way as our visual system is designed to use available information from light reflectance, our inference systems should be designed to acquire reliable information, as every increment in that capacity does translate as a survival advantage. So, again, why would humans blithely fill their minds with poor-quality information, which in most cases is of no clear advantage?

Motivated Rumors and Conspiracies

To get closer to an understanding of possible failings in our evolved quality control for beliefs, it may be of help to consider situations in which people readily acquire, but also eagerly broadcast, information of poor value. That is the case for most of the rumors that accompany salient, generally tragic events of public significance. For instance, hundreds of web pages appeared right after the 9/11 terrorist attacks in New York City, describing the probable involvement of the U.S. government or Israeli intelligence services. When the city of New Orleans was flooded as a result of Hurricane Katrina, tens of thousands of displaced persons tried to find shelter in the nearby city of Baton Rouge. It was enough that thousands of "strangers" were now in town for the stories to start circulating, focusing mostly on gruesome crimes committed by some of these newcomers. Within a few days of the refugees' arrival, most people had heard such stories, many had heard the same ones from different sources. A majority of those who heard such stories would pass them on or had already done so.[22]

To take other examples, many people in the United States are convinced that crack cocaine and the AIDS virus were engineered and spread by the secret services or other government agencies, as a way of decimating or criminalizing the black population. In other versions of the story, other communities are the target. Although no one has ever really provided evidence for these conspiracies, a great many people do believe them—and they can cite quite a few documented examples of previous mistreatment by the authorities as additional arguments.[23]

How can we make sense of all this? In psychology, the systematic study of rumors started in the United States during the war years, culminating in Gordon Allport and Leo Postman's classic *The Psychology of Rumor*. The two psychologists were specially interested in rumors that result from collective stress, in wartime of course, but also in periods of economic crisis or ecological disaster.[24] Allport and Postman explained belief in rumors as a "search after meaning," after events that are both important and ambiguous, and that trigger anxiety because of their uncertain consequences. For instance, stationed soldiers would spread rumors about impending assault to make sense of their own situation and explain otherwise mysterious movements of men and materiel. By the same token, rumors might allay anxiety, by making the world less opaque, more amenable to explanation. On the whole, the study of rumors until recently expanded and added more detail to this model, considering uncertainty, loss of control, and the need for explaining events as the main factors.[25]

But this standard understanding of rumor is not really sufficient. Everyday life in Baton Rouge does not become more meaningful once we imagine New Orleans criminals rampaging through the city. The spread of AIDS does not seem to make more sense once we assume that the HIV virus was designed by the secret services. Indeed, it would seem that the epidemic makes more sense if you do not believe the rumor. For people who believe the official account, the spread of the virus is very similar to that of other sexually transmitted diseases, and therefore not really mysterious. For those who believe the rumor, the emergence of the epidemic opens up many new, unanswered questions.

The same goes for the equally intuitive, but also equally vague, notion that rumors reduce uncertainty. It is true that floods and epidemics and terrorist attacks are unpredictable. But explaining them as political conspiracies does not seem to make them any more predictable. In fact, it adds to the uncertainty, as in both cases we have to imagine agents with great powers, whose intentions we cannot really fathom. The behavior of such daemons is less predictable than the surge of water levels after heavy rains. People who imagine that the government and the secret services engineered the 9/11 attacks are in a much worse position, when it comes to predicting and explaining such events, than those who see them as the work of Islamic terrorists.

Even if we accept that in some cases, in some sense, rumors did decrease uncertainty and anxiety (and that is far from clear), the theory still would not address the important questions about that kind of information. For one thing, why do people want to spread rumors? And people do want to transmit rumors. We generally obtain information, for example, about AIDS created by the CIA, not from reluctant sources that we implored to explain the epidemic to us but from all-too-willing and insistent individuals who seem bent on getting us to listen to them and to accept what they take to be important truths. Many people want to communicate about such things. Many of them, these days, create websites to broadcast their beliefs to the world.

Also, why does it matter to people that others believe them? People who transmit rumors are often very attentive to their listeners' reactions.

It matters to them whether one believes the information conveyed and understands its full import and implications. People who tell us that the CIA created the AIDS virus are anxious to persuade us. They and other rumormongers do not typically take divergence as a simple matter of weighing the facts of the matter. To them, skepticism and agnosticism are deeply offensive.

Rumors and Threat Detection

Rumors are about mostly negative events and their sinister explanation. They describe people intent on harming us or who have already done so. They describe situations that will lead to disaster if no action is taken. The government is involved in terrorist attacks against the population, medical authorities conspire to spread mental illness in children, ethnic others are trying to invade us, and so forth. In other words, rumors describe potential danger and the many ways in which we could all be threatened.

Are rumors successful because they are negative? Psychologists have for a long time noticed that there is what they call a "negativity bias" in many aspects of cognition. For instance, negative items in a list of words are more attention grabbing than positive or neutral ones. Negative facts are often processed more thoroughly than positive information. Or negative impressions of individuals are easier to create and more difficult to abandon than positive ones.[26] But describing a bias does not explain the phenomenon. As many psychologists have noted, one possible reason for this tendency to attend to negative stimuli may be that human minds are especially attuned to information about potential threats. That is quite clear in the cases of attentional biases. For instance, our perceptual systems work much faster and better at identifying a spider among flowers than a flower among spiders. The dangerous stimulus pops out, suggesting that specialized systems are geared to threat detection.[27]

How does an evolved mind appraise and predict potential danger? Human minds comprise specialized systems for threat detection. It is an evolutionary imperative for all complex organisms to detect potential dangers in their environment and engage in adequate precautionary behaviors.

So it is not surprising to find that human hazard-precaution systems seem to be specifically focused on such recurrent threats as predation, intrusion by strangers, contamination, contagion, social offense, and harm to offspring.[28] Humans readily attend to information about these, and by contrast tend to leave aside other kinds of threats, even if they are actually more dangerous. In the same way, children are predisposed to pay attention to information about specific threats. They are often indifferent to real sources of dangers, like guns, electricity, swimming pools, cars, and cigarettes, but their fears and fantasies are full of wolves and nonexistent predator-like monsters, confirming that threat-detection systems focus on situations of evolutionary significance. Pathologies of threat detection, like phobias, obsessive-compulsive disorder, and post-traumatic stress disorder, also focus on highly specific targets, like dangerous animals, contagion and contamination, predators and aggressive enemies, that is, threats to fitness in environments of human evolution.[29]

Threat-response systems, in humans as in other animals, face the problem that there is an important asymmetry between danger cues and safety cues. The former are actual properties of the environment. For instance, small rodents detect the smell of predators like cats and engage in appropriate behaviors. They invest more time in inspecting their environments, they avoid going across open, exposed places, they hide, and so on. This is because a specific feature of the environment—a specific smell in this case—is taken as a clear signal of potential danger. There is, however, no clear signal of nondanger. The absence of cat smell is not a reliable signal of the absence of cats, because it may occur in many situations in which cats are actually present, but their smell somehow failed to reach their potential prey.[30]

In humans, whose behavior is strongly affected by information from conspecifics, this asymmetry of threat and safety has one important consequence, that precautionary advice is rarely put to the test. Indeed, it is one of the great advantages of cultural transmission that it spares individuals from systematically testing their environments to identify sources of danger. To take a simple example, generations of tribal people in the Amazon have been told, rightly, that cassava is toxic, and that it becomes edible only after proper soaking and cooking. People do not wish to experiment on the effects of cyanide in cassava roots. Obviously, taking information on trust is a much broader phenomenon in cultural transmission—most technical know-how is handed down from generation to generation without much in the way of deliberate testing. People trust time-tested recipes, and by doing so they free ride, so to speak, on the knowledge accumulated by previous generations.[31] Precautions are special, because if we take them seriously there is no incentive to test them. If you think that raw cassava is toxic, there is nothing much to do except avoid testing the proposition that cassava is toxic.

This would suggest that threat-related information is often considered credible, at least provisionally, as a precautionary measure. The psychologist Dan Fessler tested this directly, by measuring people's acceptance of statements phrased in either negative, threat-related terms (such as "10 percent of heart-attack patients die within ten years") or positive terms ("90 percent of heart-attack patients survive for more than ten years"). Even though the statements are strictly equivalent, people place more confidence in the negatively framed ones.[32] Similarly, other studies show that people find the authors of descriptive texts, for example, about a computer program or a hiking trip, more competent and knowledgeable if the texts include threat-related information.[33]

All these factors converge to make the transmission of threat-related information more likely, which would explain why people transmit so many rumors centered on potential danger. Even not-so-serious urban legends follow this pattern, as many of them include descriptions of what may befall those who neglect potential danger. The babysitter who dried the wet puppy in the microwave, the woman who never washed her hair and unwittingly grew a colony of spiders in her hair-sprayed chignon, and other stock characters of urban legends warn us, in their macabre ways, of what happens when people fail to detect the danger posed by everyday situations and objects.[34]

So, we should expect that people are particularly eager to acquire threat-related information. Naturally, not all such information could give rise to rumors that people take more seriously than mere urban legends, otherwise cultural information would consist in nothing but precautionary advice. But several factors limit the spread of rumors about potential threats.

First, all else being equal, plausible warnings have an advantage over descriptions of highly unlikely situations. This seems straightforward, but it will impose strong constraints on communication in most cases. It is generally easier to convince our neighbors that the grocer sells rotten meat rather than that he occasionally turns into a reptile. Note that, as a matter of course, what is or is not plausible depends on the listener's own metric. Some people may be convinced by highly unlikely rumors (for example, about mysterious horsemen spreading disease and pestilence) if they have prior beliefs (for example, about the end of the world).

Second, the niche for nontested (and generally invalid) precautionary information requires that the cost of precautions be relatively moderate. To take an extreme case, it is relatively easy to convince people not to walk around a cow seven times at dawn, if they ever thought of doing such a thing, because there is no cost at all in following the prescription. In general, there is some cost, but it should not be too high. This explains why many widespread taboos or superstitions only require minor deviations from ordinary behavior. Tibetans walk on the right side of a chörten or stupa, Fang people in Gabon spill some drops of a newly opened bottle on the ground—in both cases to avoid offending dead people. Precautionary advice that is very costly will also be highly scrutinized, so that it may not be diffused to the same extent as these inexpensive prescriptions.

Third, the potential cost of noncompliance, what would happen if we failed to take precautions, should be described as serious enough that the listener's threat-detection systems are activated. If you are told that the only consequence of passing on the left side of a stupa is that you may sneeze, you will probably ignore the rule. Offending ancestors or deities seems far more serious, especially if it is not clear precisely how they might react to the insult.

So, it would seem that threat detection is one of the domains in which we may have to turn down our epistemic vigilance mechanisms and take as a guide to behavior precautionary information, especially if it is not too costly to follow, and if the averted danger is both serious and uncertain.

Why Threat Is Moralized

When we consider junk culture, it is easy to remain fixated on the question, Why do (other) people believe such things? Why not ask a question that is just as crucial, namely, Why do people want to transmit such information? Why tell each other about penis thieves and the intelligence services' role in creating the HIV epidemic? Questions about belief are fascinating, but they may not be as important for cultural transmission as we would like to think. True, many people believe the rumors they propagate, but belief is not sufficient. There is also a motivation to transmit, without which many people would cultivate their own poor-value information, but there would be no rumors, no junk culture.

In many situations, the transmission of low-value information is associated with strong emotions. People consider information about viruses and vaccinations and government conspiracies as terribly important. When they transmit information about such topics, people are not just eager to convey but also eager to convince. They do pay attention to their audience's reactions, and they consider skepticism highly offensive. Doubt is attributed to all sorts of wicked motives.

Consider, for instance, the campaigns against children's MMR (measles-mumps-rubella) vaccination, movements that began appearing in Britain and the United States in the 1990s. People who spread information about the dangers of the vaccine, which in their view may have caused autism in previously normal children, did not just describe the alleged dangers of the vaccines. They also vilified physicians and biologists whose research was at variance with the autism-vaccination theory. They described doctors who administered the shots as monsters who knew perfectly well that they were endangering the lives of children but would rather get money from pharmaceutical companies than stand up to them and tell the truth.[35] If you agree with most physicians, that the minor side effects of mass vaccination are worth the collective protection they afford, you are siding with criminals.

The same goes for many other instances of widespread rumors and conspiracy theories. The release of HIV infection in the population, according to the rumors, was not an accident. It was part of a deliberate plot to kill

Africans (or black Americans in some versions). The continued silence of the authorities on the matter is not ignorance or incompetence, it is evidence of a cover-up. And, just as in the case of vaccination, the listener's reaction is moralized too. If you express doubts that the government would devise some elaborate plan to kill as many civilians as possible, in collaboration with the secret services, or if you merely point out that the evidence is not altogether compelling, you may be denounced as just another shill of the government, a supporter of biological warfare against the citizenry.

Why are the beliefs so intensely moralized? One obvious answer is that the moral value of broadcasting the information, and of accepting it, is a straightforward consequence of the information conveyed. If you really think that the government has tried to exterminate some ethnic groups or helped plan terrorist attacks on its citizens, or that doctors are deliberately poisoning children with vaccines, should you not try to make that known, and to rally as many people as possible?

But this may be one of the apparent self-evident explanations that just raises more questions than it solves. For one thing, the connection between belief and the need to convince others may not be as straightforward as we commonly imagine. The social psychologist Leon Festinger became famous for studying millennial cults, whose predictions for the timing of the end of the world have clearly failed, and for observing that this clearly refuted belief led members of the group to more, not less, proselytizing.[36] Why would that be the case? Festinger's own explanation was that minds strove to avoid cognitive dissonance, a tension between incompatible beliefs, for example, that the prophet was right and that his predictions had failed.[37] But that was not entirely satisfactory. Indeed, it failed to explain one major aspect of the study of millennial cults, that failures in prediction had led people not just to concoct excuses for the failure (which would indeed minimize dissonance) but also to recruit more members for the group. The effects of the alleged dissonance were mostly seen in people's interaction with outsiders, and that requires an explanation.[38]

It may help to step back and consider all this from a functional standpoint, considering mental systems and motivations as designed to solve adaptive problems. From that standpoint, it is not clear why human minds would try to avoid cognitive dissonance, when the discrepancy between observed reality and one's prior beliefs is an important piece of information. Going further, we should ask why the reaction to an obvious failure would be to reach out and try to get more people onboard.

The process makes more sense if we see it in terms of the coalitional and group-support processes I described in chapter 1. Humans need social support, and they need to recruit other individuals to join in collective actions of various kinds, without which there is no individual survival. A crucial part of our evolved psychology consists in capacities and motivations for efficient coalition management. So, when humans convey information that may persuade others to engage in specific actions, we should try to understand this in terms of coalitional recruitment. That is to say, we should expect that an important part of the motivation here is indeed to persuade others to join in some collective action.

This is why the moralization of opinion might seem intuitively appropriate to many people. Indeed, evolutionary psychologists like Rob Kurzban and Peter DeScioli, as well as John Tooby and Leda Cosmides, have argued that in many situations moral intuitions and feelings are best understood in terms of support and recruitment.[39] The arguments and evidence are complex, but the central point is straightforward and clearly relevant to rumor dynamics. As Kurzban and DeScioli point out, for each moral violation there is a transgressor and a victim, but also third parties, people who condemn or condone the behavior, protect the victim, impose reparation or punishment, withhold cooperation, and so on.[40] It is in these people's interest to side with the party more likely to attract other supporters. For instance, if one individual takes more than her share of the communal meal, each bystander's decision to ignore or punish the freeloader is influenced by his representation of all the others' reactions. Now moral intuitions about the relative wrongness of different behaviors are automatic and largely shared among human beings. In other words, each agent can predict the other's reactions from her own emotional response. As people can expect this rough consensus, it follows that a moralized description of a situation is likely to result in coordinated opinion, more so than other possible understandings of what is going on. People tend to condemn the party they see as the transgressor and side with the victim, partly because that is also the choice they expect others to make.[41]

From this perspective, the moralization of other people's behavior is an excellent instrument for social coordination, which is required for collective action. Roughly speaking, stating that someone's behavior is morally repugnant creates consensus more easily than claiming that the behavior results from incompetence. The latter could invite discussions of evidence and performance, more likely to dilute consensus than to strengthen it.

This would suggest that our commonsense story about moral panics may be misguided, or at least terribly incomplete. It is not, or not just, that people have beliefs about horrible misdeeds and deduce that they need to mobilize others to stop them. Another factor may be that many people intuitively (and of course unconsciously) select beliefs that will have that recruitment potential, because of their moralizing content. So the millennial cults with failed prophecies are only a limiting case of the more general phenomenon whereby the motivation to recruit is an important factor in people's processing of their beliefs. That is to say, beliefs are preselected in an intuitive manner, and those that could not trigger recruitment are simply not considered intuitive and compelling.[42]

We should not take this speculative explanation to suggest that people who spread rumors are cynical manipulators. They are in most cases unaware of the mental processes that make a moralized description of behavior highly salient to them and others, salient in the same manner, and most likely to attract support for the cause. Because we evolved as support seekers, and therefore recruitment specialists, we can orient our behavior toward more efficient coordination with others without having to be aware of it. Also, we should not imply that such appeals to morality are invariably successful. Moralization can boost recruitment, but it does not guarantee it.

The Template for Crusades

A whole variety of social movements involve such recruitment dynamics, in which some agents broadcast information that is particularly effective at persuading others to join. That is the case in witchcraft accusations, for instance, when an individual claims to have suffered magical attacks by a relative or some other resident in the village and enlists the support of the rest of the community to make the alleged witch confess his sins and atone for his malfeasance.[43] The dynamics of recruitment for a cause—for a crusade, to use a convenient term—is one where an individual or small group hits on the kind of information that persuades many others to join the coalition, often at great personal expense or peril.

People who call for a crusade—to abolish drinking, to riot against the government, to burn down an ethnic neighborhood, to defend children against vaccination, or for any other purpose—are broadcasting a very special signal to their audience. First, they are generally focusing people's attention on some (real or imagined) threat, which they describe as potentially harmful not just to themselves but, crucially, to many others as well. Second, they suggest that dealing with that threat is a matter of collective action, that help is needed. Third, they describe participation in that collective action as a moral imperative. Acting for or against the proposed cause is no longer a matter of knowledge or acceptable preferences—it can be seen as an index of moral character.[44]

This interpretation of rumors and other such emotion-laden communication may perhaps seem overstated. I suggested that the main motivation here is recruitment toward collective action. But in many cases there is no precisely planned action. The websites that denounce official involvement in the 9/11 attacks do not (generally) ask people to overthrow the government, and people who claim that a racist CIA spread the HIV virus (generally) do not call on black people to attack their local CIA office. The strategic value of calls to action remains the same, however, even if no specific collective action is proposed. This is because people need signals of potential support, long before actual support is needed. So they need to be able to evaluate alliances and solidarity around them. As we saw in the previous chapters, there is experimental evidence that people do monitor the social environment, and that they automatically detect alliances on the basis of indirect or implicit information—who knows whom, who sided with whom in the past, which people have a similar accent, and other such cues.

Broadcasting threat information may have the effect of forcing people to provide such information. That is, people who agree with information you provide, precisely because you provided it, signal that they are ready to follow your cause, to join some collective action that you may instigate. By contrast, those who ask for evidence, or debate the plausibility of your claims, signal that any solidarity with you would be conditional, which is of course not what we want of coalitional allies. When the people of Baton Rouge circulate frightening stories about New Orleans refugees, they confirm to each other that they are part of the same group, with common interests they do not necessarily share with strangers and newcomers, and they possibly test each other's commitment to the community.

Moralized recruitment of this kind can lead to competitive outrage. If your reaction to threat information is an index of your moral value, and of your commitment to potential collective action, then you are motivated to make that reaction clearly visible to all. But in a group where everyone believes in the rumor, and everyone is morally offended, a clear signal of commitment is to be more outraged than most. This dynamic is present in many militant movements, in which people gain acceptance or even authority by claiming that the situation is not just undesirable but also unacceptable. This naturally leads to a parallel escalation in the actions proposed. When there is some uncertainty about people's commitment to the cause, it becomes necessary for many members of the coalition to demonstrate their resolution by adopting or recommending stronger action and more extreme positions than other members of the group. If others agree that we should ban hard liquor, advocate for a ban on all alcohol. If others agree that the group should shun renegades, argue that we should assault or kill them. That drift toward the extremes is also visible in movements with fierce competition for leadership, as the would-be head of the movement cannot be seen to be less committed than anybody else. This dynamic was observed in movements as diverse as the IRA in Ulster, the Tamil Tigers of Sri Lanka, the American Ku Klux Klan, and the Peruvian terrorists of the Shining Path.[45]

Truth-Making Institutions

What is the antidote? It would seem that many factors converge to turn some forms of junk culture into large-scale epidemics. Threat-related information is processed in particular ways in the mind, in some cases bypassing our epistemic vigilance. Moralized threats are powerful recruitment signals that can be used for the building of coalitions, an evolutionary imperative. So we should not be surprised that low-value information seems so pervasive and persuasive.

That is not the whole story, however. Humans also created many kinds of epistemic institutions, that is, sets of norms and procedures supposed to guarantee the production of true information. These attempts are not always successful, but they reveal a motivation to seek reliable information, through institutions that expand on our spontaneous epistemic vigilance and extend its scope. Here are several illustrations.

There are divination procedures in almost every known human group. People trust the flight of birds or the throw of dice to provide information that is, to some extent, believed to be more certain than ordinary statements and opinions. Most divination is not so much about the future as about unobservable states of affairs. Are the ancestors angry with you? Are your in-laws jealous of your success? Is your spouse or business partner really committed? And so on. In all these domains, what diviners provide is something different from expert opinion or wisdom. It consists in some technique that supposedly offers a guarantee of truth. Why would anyone resort to such techniques? Skepticism about divination is not a modern phenomenon—Cicero expressed it in the most forceful terms.[46] But there is a market for such guarantees, hence the success of divination and mediumship, including in modern industrial societies where official knowledge institutions deride or despise divination.

What makes divination compelling? Why would the entrails of sacrificed animals provide information that is not available from competent and wise people? Why would a pack of cards say more important things about your family than a well-meaning friend? In other words, why bother with the procedure at all? The answer may be that divination procedures result in a diagnosis that is construed as, precisely, not formulated by any human agent. Indeed, a constant claim in divination rituals is that neither the diviner nor her audience has any effect on the diagnosis. The cards are shuffled so no one could guess which ones will come up and in what order. The dice are thrown rather than placed on the mat, which makes it impossible to guess what numbers will come up. The birds in the sky cannot be controlled, and the sheep's entrails were not visible before the sacrifice. In other words, whatever the divination "says," it is apparently not said by the diviner. The reality is often very different, of course, but this emphasis is telling. From pure mechanical divination (such as throwing dice) to inspired prophecy, the agents who formulate diagnoses are explicitly presented as not the source of truth. The main assumption of practitioners and clients is that, given a certain reality, the diagnosis could not have been otherwise, strongly suggesting that the diagnosis, the way the dice rolled or the way the birds flew, was actually caused by the situation it describes.[47]

This apparent impartiality of random procedures is why, in some contexts, divination may seem more efficient than the available alternatives. In Liberia, people sometimes use the sassywood ordeal to establish a defendant's guilt. People suspected of murder drink a decoction of sassywood leaves, a potent poison that, according to the theory, will promptly kill them if they are guilty. If a presumed perpetrator refuses to submit to the ordeal, this of course amounts to an admission of guilt. This poison ordeal is certainly not an optimal process of discovery. It is true that the poison sometimes kills and sometimes not, but that of course has little to do with a defendant's guilt. Still, compared to other procedures available in the region, the process is at least somewhat impartial, which makes it, according to some legal scholars, clearly more efficient than the expensive and largely corrupt official justice system in Liberia. Sassywood ordeals are accessible to all and provide at a low cost some nonrandom (if not entirely reliable) information about guilt.[48]

This leads to another salient domain of epistemic institutions, that of legal argument. In very different cultural environments, people have established rules of evidence and inference, supposed to guide judges in establishing evidence and assigning responsibility and guilt. This is obviously not a monopoly of Western legal traditions, and complex sets of norms of this kind can be found, for instance, in the classical Chinese legal system.[49] There are also examples of these kinds of norms in some nonliterate cultures. For example, Trobriand islanders developed a complex system of norms and arguments to do with land tenure and the adjudication of competing claims in that domain.[50]

A third and most salient example is the development of scientific institutions, and more generally the very improbable kind of social interaction that created scientific research as we know it. Decades of social science research on science have demonstrated that science is not produced by isolated minds, and therefore we require a highly specific social organization of science, so to speak.[51] But it has been difficult to specify what is special about that particular social interaction, that is, why it produces knowledge of higher accuracy, precision, and explanatory power than any other human endeavor. Perhaps that will be illuminated by taking into account the kinds of cognitive capacities and evolved motivations activated in the context of scientific activity, a project that is only in its infancy.[52]

The existence of truth-validating institutions would suggest that our epistemic future is not entirely bleak, why human societies are not necessarily doomed to sink ever deeper in an ocean of misinformation. Against this optimistic conjecture, one could argue that technology, especially the availability of Internet-based communication, should enhance the diffusion of junk culture, for two reasons.

First, as we all know, connectivity makes it cheap to acquire information and makes the cost of broadcasting almost negligible. That is not just because connections are cheap but also because the role of reputation is greatly diminished. Again, consider a small-scale society. In such a group, accusations of witchcraft, for instance, are potentially very costly. You never know for sure that people will not rally around the alleged witch. You may pay dearly if you are the only one to level the charge against a particular individual. That is why public accusations of this kind only occur after a long period of discreet consultations, and in some places are never made public.[53] By contrast, modern connectivity allows both anonymity and geographical distance, virtually eliminating the social costs of accusations and rumormongering. So it is no surprise that Internet rumors and crusades are so vicious in tone, so quick to emerge, and often so expansive.[54]

Second, worldwide connectivity may fuel our worst dispositions to create and broadcast junk culture, by fooling us into illusions of consensus.

Consider this. In a small-scale society or in a village, those who come up with some new variety of, say, magical beliefs, will probably find very few people who share their strange notions. By contrast, in a connected world that includes billions of users, almost any kind of harebrained proposition is probably already promoted by thousands of individuals or more. So connectivity is likely to provide all with an inflated sense of consensus around their own ideas—a propensity that was already observed in experimental studies.[55] This effect could be multiplied by the illusion that sources are independent. That is to say, if we find out that thousands agree with us, for example, that the world is indeed controlled by alien reptiles, we marvel at the fact that so many great minds think alike. It would seem that all these individuals independently converged on that same theory, when in all likelihood many of them read the exact same web page.

In the face of these factors, perhaps one counterstrategy would be to try to make truth useful. If there is an immense market for information from others, there must be a very large market for organizations that supply some control over the quality of that information. That is, providing evaluations of other people's claims to truth may become a desirable service, which can be remunerated in money or reputation. One may object that this could end up in an endless regress, as the guarantors must themselves be guaranteed by metaguarantors, and so forth ad infinitum. That is certainly the case, if one is looking for an ironclad guarantee of epistemic quality. But that is not really required. What is needed is a good enough guarantee of validity, strong enough to make the cost of misinformation, at least in some contexts, too high to be worth the miscreants' time and effort.

True or Useful: What Is Information For?

The domain of junk culture is vast and many-splendored. I emphasized threat detection, rumors, and moral crusades, three overlapping phenomena, because of their social impact. They create social dynamics that can mobilize very large numbers of people, coordinate their actions in spectacular ways, and lead to massive social and political change. Crusades, beneficial or not, spread information that serves recruitment. In that sense, they are useful to the participants, either moral entrepreneurs or mere followers, as they provide shared representations that make coalitional alignment, and collective action, possible.

African penis thieves and the great wave of English Satanic abuse exist only in the imagination of rumormongers and rumor believers. So belief in such occurrences seems to challenge our notion that minds were designed to acquire and produce useful information about environments. Why do we sometimes misbelieve? Ryan McKay and Daniel Dennett, a psychologist and a philosopher, proposed to consider seriously what they called "adaptive misbelief," that is, situations in which epistemic mishaps, or to be blunt, erroneous beliefs, may be advantageous to people who hold them in terms of fitness. The diffusion of low-quality information, in rumors, legends, and conspiracy theories, may seem an example of this process. The beliefs are certainly not useful, if we consider one major function of mental systems, for maintaining an accurate and useable representation of the organism's environment. Mental systems are not, however, constructed to be useful only in this sense; they are designed, more generally, to enhance fitness. That does include, most of the time, that the representations produced are indeed accurate. But that is, precisely, most of the time and not all the time.

We generally assume that information is transmitted because of its epistemic value, its connection to the way things are and to potential consequences for fitness. That explains the transmission of vast domains of cultural knowledge, but also of deceptive communication, which favors the deceiver's interests precisely because it is false. But epistemic value is not the only factor that motivates humans to spread information. The need to be seen as a reliable source, the requirement to detect threat information, the urge to recruit others in collective action, or at least to gauge their potential commitment, are powerful factors. As they are not directly affected by the value of the information transmitted, junk culture is in some conditions both epistemically disastrous and evolutionarily advantageous.

Why Are There Religions?

... And Why Are They Such a Recent Thing?

SURELY, ONE EXPECTS, EVERY HUMAN society has a religion. Or several. Early travelers and explorers shared that expectation, and so did early anthropologists. To European scholars, it was obvious that there would be religions in all those exotic bands, tribes, kingdoms, and empires. But things turned out to be much more complicated than that.

True, most empires or states seemed to have something approaching a religion, with priests, codified ceremonies, and most important, a set of doctrines including eschatology, theodicy, and soteriology, or in plain English, about the order of the cosmos and the end of our world, the origins of evil and the path to salvation, respectively. That was not too surprising, as Europeans long before anthropologists had been acquainted with the civilizations of India, China, and the Arab world.

But one could not find any of those things in the smaller-scale societies. There seemed to be no religious organizations, in the sense of a caste or group of specialists of supernatural affairs, with formal training and a certification process. More puzzling, people in those societies seemed to have no interest whatsoever in the salvation of the soul or the ultimate origin of evil. Most frustratingly for anthropologists, they generally seemed to have no established set of religious beliefs. True, they would talk about ancestors, souls, and spirits. But these statements would often be vague, idiosyncratic, or even inconsistent. Whatever those people had, it did not seem to fit the expected picture of a religion. Early anthropologists trying to describe the "religions" of small-scale societies found that they were very much trying to fit square pegs into round holes.

In some quarters, this led to interminable and in fact intractable ter-
minological debates about the proper definition of the term "religion."
These did not help answer the questions that really matter, such as, How do
human minds represent religious concepts and norms? Are there similarities or even universals in this domain? Is having such representations a
natural consequence of the way our minds work? Is that a consequence of
natural selection? Is it even adaptive?

Supernatural Combinations

The proper place to start, in order to understand the various things called
religion, is in the human capacity to entertain supernatural fantasy. This
vast domain of cognition includes daydreaming, fiction, myth, dreams, all
produced by what classical psychology would have called the faculty of
imagination.[1] Of particular interest here is what could be called the "supernatural" domain, those imagined entities or beings or processes that do not
belong to our natural world—indeed, are in many cases excluded by natural
laws. For example, ghosts, like people, can walk around and apparently perceive and understand what is going on around them, but they differ from
people in that they are actually dead and can pass through physical obstacles like walls and closed doors. Zombies are animated corpses that usually
feed on human flesh, whose behavior is not controlled by their own, vacant
minds. Vampires are immortal ex-humans who feed on human blood.
Golems are made out of clay and turned into animate beings by a magic
formula. Supernatural imagination is not confined to human-like figures of
this kind. Consider the talking animals of so many folktales, the mushrooms
turned into humans as in Ovid, and a man into a donkey as in Apuleius. Not
to forget amulets that move around of their own accord, or statues that
bleed or cry or otherwise convey their emotions to humans.

A limited number of underlying themes are found again and again in
this vast and varied repertoire of oddities. Particularly frequent is the implied violation of some very general expectations about physical objects,
living things, animals, and so on. Humans tacitly assume that the shape
and behavior of animals are limited by their inherited essential traits, but

metamorphoses contradict that; we assume that animate beings behave because of their own intentions and goals, but zombies show the opposite; we
assume that man-made objects are inert and inanimate, but some statues of
Ganesha drink offerings of milk. Even extraterrestrial visitors are described
in that particular manner, as combining surprising powers and human-like
intentions.[2] Supernatural notions of this kind pop up in highly similar
forms the world over. In the most diverse cultures, one can find notions of
corpses rising from the grave, talking animals, dead souls coming back to
haunt the living, or magical objects that become animate or respond to
incantations.

It may seem paradoxical that the domain of fantasy, and supernatural
imagination, turns out to be largely repetitive. But that is because imagination does not create ex nihilo, it just recombines preexisting conceptual
material. Indeed, the exercise of fantasy is of great interest for cognitive scientists because it reveals implicit or unconscious principles that inform everyday cognition, as Kant pointed out (after many others, no doubt).[3] For
instance, when people are asked to imagine entirely novel animals, they
construct beings with left-right (but not top-bottom) symmetry, and with a
preferred direction of motion with facing sense organs. This suggests that
implicit principles, derived from some actual animals, govern imagination
in the domain of fantasy.[4] That is also true of imagined beings in fiction and
myth, even in genres where people are precisely trying to push the boundaries of the imaginable, like science fiction and fantasy literature.

For all the fanciful details, many supernatural notions are constructed
in a very simple manner, generally combining two distinct kinds of elements. One is a salient, explicit violation of our expectations, like the fact
that a person can go through walls or a statue can drink milk. The other
ingredient is a whole set of expectations that do apply to the imagined entity, like the fact that the ghost has perception, memory, and intentions, or
that the magical statue is made of wood, has weight, is only in one place at a
time. The part that violates our intuitions is the one we generally focus on,
and is described explicitly. By contrast, the part that confirms our intentions is generally left implicit—it goes without saying and is generally left
unsaid.[5]

To be more precise, supernatural concepts combine salient violations and implicit confirmation of what are called intuitive ontologies, that is, sets of expectations that we entertain about large domains of reality, such as animate beings, persons, living things, man-made objects, natural things.[6] For instance, specialized inference systems detect some kinds of motion as typical of animate beings, which triggers an automatic representation of their motion as directed by internal goals. In a different domain, when an object is identified as man-made, other inference systems try to identify their function. The behavior of people around us triggers inference systems that produce some picture of their possible intentions and beliefs. All of this happens automatically and for a large part unconsciously—all we are aware of are the results of these computations.[7]

Supernatural notions include an explicit violation of the expectations produced by these inference systems, which makes them attention grabbing. A person goes through walls and a statue is drinking. But a great deal of what makes these notions useable, so to speak, is the work of those intuitive expectations that are not violated. The ghost still remembers, and the statue is in a particular place. If we did not have these background assumptions, supernatural notions could quickly become useless—consider a ghost with no memories or a statue that is nowhere in particular. Indeed, a host of experimental studies have shown that such combinations of limited counterintuitive materials and massive preservation of all other intuitive expectations are particularly salient and usually better recalled than other conceptual combinations.[8]

This suggests that, all else being equal, such concepts will spread better than other possible variants, that is, become what we call cultural representations, which people in a group entertain in a roughly similar way. By the same token, the connection between inference systems and the supernatural imagination also explains why the latter is so similar the world over. Supernatural fancy is based on simple and limited tweaks of intuitive expectations that are part of our evolved mental design and the tweaks therefore occur in a markedly similar way in all human minds. Anthropologists and historians used to think that the notion of a "supernatural" domain was certainly a culturally specific construction, as it seemed to require a specific notion of "nature," against which one could maintain that flying trees and drinking statues were in some sense beyond the natural. Now, only in very few human cultures, like China and ancient Greece, do people engage in systematic, explicit reflections on what nature is, on what its general laws amount to.[9] But the point made here is not about explicit reflective thoughts on nature but about intuitive understandings of the physical and social environments. What makes a talking tree or an invisible goat attention grabbing are not explicit notions about what is natural and what isn't but entrenched intuitive expectations about physical objects and biological processes that are typical of human cognition.[10]

Religious Traditions

The supernatural menagerie is certainly entertaining and deserves study, as it reveals implicit, entrenched conceptual principles. But one reason it has attracted the attention of anthropologists and historians is that most concepts usually described as religious belong to this broad domain. A restricted subset of supernatural notions become formalized, codified, in what have been variously called religious traditions, cults, primitive religion, and so forth.

A few examples may help here. Consider, for instance, the ancestor cults found in virtually all small-scale societies, especially so in agrarian and pastoralist societies. People engage in imagined interaction with their ancestors, who are said to demand occasional sacrifices as well as some measure of respect for traditional practices and mores. Dead people are said to hang around as disembodied souls, who remain in a liminal world between the living and the dead, and cause much mischief, notably because of their longing for the world they left. The transition from unstable, spooky ghostly presence to stable and conservative ancestor is a frequent rite of passage in such societies, which is frequently the point of the second funerals performed in so many places the world over, during which the remains of the dead are often given a final resting place, marking their installation as ancestors.[11]

Another set of traditions center on what anthropologists call shamanism or mediumship, interaction with specific souls or spirits that is supposed to remedy some misfortune. Illness, but also disappointing crops, weak livestock, or social strife, may be interpreted as caused by angry souls or ancestors, or as requiring the help of some of these superhuman agents. In such cases, contact and negotiation with superhuman agency requires ceremonies and in most cases the participation of specialists such as mediums, healers, shamans, diviners—all of whom in some way or other are thought to be especially qualified to figure out what superhuman agents want, or how best to placate them.

Such cults and traditions are almost universally about human-like agents, with specific superhuman features. This stands in contrast to the baroque variety of supernatural imagination. Human minds are attracted to notions of floating islands, talking trees, or firebirds, but the cults and traditions are almost invariably about agents. Ghosts, ancestors, souls, spirits are described in terms of salient, explicitly counterintuitive properties, like being in several places at once, remaining invisible, having complete knowledge of what people are up to, and so on. But they are also tacitly understood as having minds, mostly similar to human minds. They perceive, think, feel, and remember in ways that seem very natural to us because they are assumed to be human-like.

As they are a subset of the supernatural, concepts of gods and spirits and ancestors inherit many important properties of the broader domain. One salient feature is the contrast between the counterintuitive and intuitive components of these concepts. The former is generally explicit, or accessible to conscious inspection, and grabs people's attention. The latter component need not be represented explicitly, and in general is not. That is, people do not need to remind themselves that "the goddess, having a mind, remembers what happened to her," because that literally goes without saying, as an extension of our everyday intuitive psychology to an imagined being.

For people used to modern, institutional religions, the most surprising aspect of these practices or traditions is that they do not come with a doctrine, a stable set of assumptions that most participants could readily explicate and would agree with. For instance, people may feel that it is crucial to give the ancestors their ritual due, in terms of sacrificed animals, but they do not have a clear notion of why ancestors need this, how they profit from the sacrifice, how they would find out if people did not oblige, and so forth. This may seem surprising to outsiders—How could you take the ritual seriously, yet not have a precise notion of what it does?—and, at the same time, all too familiar to anthropologists who have worked on such traditions and know from bitter experience that ethnography is difficult. In most places you cannot just ask people "What are ancestors like? How are they different from you and me?" without being met with baffled incomprehension. Most people never thought about such things, and those who did came up with the most extravagant, idiosyncratic elaborations. The task of an anthropologist is to sieve all this information and formulate a set of minimal assumptions about, say, ancestors that seem to be clearly implied by most statements and practices about them, such as that they are not usually visible, that they demand pigs not oxen, that they are forgiving or vengeful, and so on. It is a difficult task, because there is no stable explicit doctrine at all.

Another surprising aspect of these traditions is that there is no organization of religious specialists, that is, no equivalent of a priesthood, a caste of Brahmin ritual officers or a group of religious scholars like the 'ulema. There is no formal training, no school for mediums or shaman-training programs, that people should attend to be considered legitimate practitioners. Instead, such specialists are often said to have undergone a special initiation, and in most cases they learned their craft as apprentices to an established specialist. But these are neither necessary and sufficient nor necessary conditions for dealing with ancestors.

In many places, this special skill is considered to be an essential, natural property of such people. For instance, in Cameroon the *ngengang* specialists are said to have a special extra organ, *evur,* that allows them to deal with ghosts and souls. It is of course uncertain whether any given persons do or do not possess this extra organ—only by their deeds (or rather, their successes) can ye know them, so to speak.[12] This notion of an extra organ is widespread in central and southern Africa. In other places, people do not reify it as an organ, though they still assume that a skill at handling spirits requires a special, natural, intrinsic quality that sets the shaman apart from the regular folk. In the Tuva Republic of southern Siberia, for instance, people attribute the shaman's results to possession of some inner "force" that shamans are born with and carry for the rest of their lives. Indeed, in Tuva people consider each shaman to have his or her own essence that makes the specialist unique and uniquely gifted to deal with spirits. Here, too, there is a great deal of uncertainty as to whether a given specialist is or is not the real thing. People readily admit that many shamans are impostors, or simply incapable. But, again, only experience can tell.[13]

Finally, the most striking difference between such religious crafts and what we commonly expect from religion is the absence of a notion of faith, or of a community of believers. Anthropologists have long argued that notions of faith or belief, familiar from modern forms of organized religion, do not travel very well—they seem entirely irrelevant in the description of less familiar religious traditions.[14] This also extends to the notion of faith as commitment. People who resort to a diviner or shaman are not joining a cult, in the modern sense. They are just recruiting a specialist, as one would a carpenter or plumber. Making use of the local shaman does not make you a member of a particular community, any more than going to the dentist's would make someone a member of the congregation of dentistry believers. The same point applies to ancestor cults, the other main form of religious tradition found in small-scale societies. People who sacrifice a pig to their ancestor are not joining a group—they already are part of a lineage, which is precisely why they want to exchange favors with one particular ancestor. Instead of faith or commitment to deities, what we find is an empirical bet, the expectation and hope that particular rituals will placate the ancestors, that this particular shaman will help heal this particular patient, and so on.

Perhaps the most salient difference between religious traditions of small-scale societies and what most modern people are familiar with is this intensely pragmatic concern. In such places, people do not engage in religious activities to understand the cosmos, figure out the meaning of their existence, or understand the foundations of morality. They simply want their crops to thrive and misfortune to fall far away. Once these practical concerns are addressed, people in such societies are generally not interested in metaphysical questions.

Such are, in very broad strokes, the practices and cults that we can observe throughout human history, all over the world. There are many differences, of course. Foragers, for instance, generally have much less interest in ancestors, spirits, and witches than agrarian populations. And in many large-scale, complex societies the cults are in competition with more organized forms of religious activity, as we shall see presently. But it is worth emphasizing that for most of human history, and in many human societies to this day, interacting with superhuman agents was and is being done without an established, explicit doctrine, without formal training for specialists, certainly without any notion of belonging to a community of believers, and mostly for practical reasons.

Religions and the Invention of "Religion"

With the advent of large-scale kingdoms, city-states, and empires, specialization and division of labor became much greater than in less stratified tribes and chiefdoms. This was most remarkable in the development of craftsmen's lineages or castes, in the emergence of tradesmen as a class, and, most fundamental, in the existence of a distinct group of functionaries, soldiers, bureaucrats, scribes, enforcers, and defenders of the centralized power. This is also the period when a distinct group of ritual specialists appeared, the people we call priests, with exclusive rights to some ritual activities, particularly those closely connected to political power. Their influence gradually extends beyond that, and in many cases they reached a position of quasi-monopoly in the provision of ritual services. This development is in some ways continuous with the cults and traditions described above. The activities in question are supposedly addressed to superhuman agents and include ritualized behavior as well as routinized ceremonies somewhat similar to those found in smaller-scale societies. But there the similarity ends, as many features are unique to this new social and political phenomenon.

The most important development is that there is now a distinct organization in charge of interaction with superhuman agency. As I mentioned above, ancestor cults required participants in specific positions, for example, an elder to perform a sacrifice, and shamanistic performances were carried out by specialists. But in neither case was there any organization in charge of such ritual activities. The castes and groups of priests constitute such organizations. That is, questions such as who should perform what, when, and in what manner are now handled by collectives of specialists, some extremely hierarchical, others fairly egalitarian, and as a result they are now dealt with in a fairly uniform manner within a city-state or a kingdom.

This stands in contrast to previous religious cults. Ancestor ceremonies, for instance, were performed according to what elders could recall of past, and therefore supposedly proper, performance. Shamans and mediums improvised and often created their own variants of ritual sequences. All this is replaced with a unique set of procedures agreed among the priests. Literacy, where available, increases uniformity by making it possible to codify procedures, for example, to stipulate the proper sequence of ritual performance, the amounts of offerings required, and so on. With the spread of literacy, there appear written doctrines and religious regulations, which makes it easier for priests to offer regulated, standardized rituals, based on stable recipes.[15]

The rituals provided by one member of the priestly group could be provided, at least in principle, by any other such agent. Also, the criteria that make one person a valid agent of the organization are now codified, and they apply to all members in the same way. Mystical aspects of training, such as initiation and visions, are typically replaced with a form of training, like memorization of texts, or literate competence, that can easily be verified by the leadership. No priest will claim that, having received direct divine inspiration, he can dispense with that formal training and induction into the organization.

Most religious organizations of this kind become closely linked to centralized political power. Political influence is clearly necessary to any group that gradually extends its influence, and it replaces all sorts of informally provided services with a unique set of religious goods. That is why, in the history of city-states and empires, the priests are forever trying to increase their political clout. There is nothing Machiavellian in this description of the organizations' political action. It does not even imply that the priests deliberately seek political influence—it is just that whatever organizations did not proceed in this manner did not actually survive.

Most important, the religious notions themselves are of a new kind. Shamans dealt with souls, ghosts, or spirits; ancestor cults were about the lineage's forebears. All these cults were about local superhuman agents, tied to a particular place or social group. But organized priests typically describe themselves as interacting with more universal agents, gods whose jurisdiction extends to the entire polity, be it a city-state or an empire. Gods are described as being ubiquitous and as having unlimited powers and perception or complete prescience.[16]

As priests offer standardized ways of interacting with particular gods, there appear explicit, and fairly uniform descriptions of the gods in question. As the anthropologist Harvey Whitehouse puts it, there is a stark contrast between such doctrinal developments and the mostly "imagistic" practices of shamans and mediums, in which concepts of superhuman agents are either not explained at all or mentioned during rare, exciting, and conceptually ambiguous rites. In many cases, the doctrines of organized religions become internally consistent, explanatory, deductive. They can be distilled into lessons for the masses and more elaborate versions for specialists.[17]

Such religions first appeared in places like Egypt, Sumer, the Mayan and Aztec empires, India, and China. In all these places, specialist priests became the central, officially sanctioned providers of ceremonial service. Groups of priests took over some of the rituals previously performed by shamans or local specialists. The notions of local gods and spirits were replaced with doctrines about cosmic gods. The castes or lineages of priests were closely allied to the royal court, or in some cases constituted the royal entourage.[18]

Note that some important features of what modern people would associate with the term "religion" are absent from this picture. People in such places do not choose to join in a particular cult—they are enlisted to service the gods, erect temples, and contribute to sacrifices. There is no clear notion of personal "faith" as relevant to religious activities—one must obey the gods and priests, and abide by the doctrine's prescriptions and taboos. Most important, the gods are not described as interested in personal morality, and they certainly do not behave as moral exemplars.[19]

It is at this point of social development that people can start using an explicit concept, "religion," to denote a special domain of human thought and activity. "Religion"—or rather, of course, the various local terms used in these different societies—is the name of whatever organized ritual specialists deliver, which usually combines a stabilized set of official beliefs, with specific prescriptions and taboos, specific rituals, and the training of a specialized personnel. In a similar way, people before intensive division of labor did not think that there was such a thing as "craftsmanship." For foragers or horticulturalists who make their own tools, weapons, clothes, and toys, the term would have little meaning. In the same way, people in the small-scale societies in which we evolved generally have no specific term for what we call politics or religion. It is when a special group of people specialize in a range of goods or activities, like religious services, that a special notion "religion" makes sense.

Religions Win Battles, Lose the War

It would be wrong to suggest that each form of religious activity simply replaced the previous ones. But the emergence of religious organizations did not result in the elimination of ancestor worship, mystical cults, or shamanistic practices. First, not all human groups turned into large-scale unified polities with extensive social stratification and division of labor. Also, religious organizations in most places, for most of human history, had to contend with the constant presence of shamans and mediums, and other similar, nonorganized religious service providers. Indeed, a good deal of the history of religions is the history of their fight against these competitors. As noted above, organized groups of priests naturally strive toward monopolization, and in the same process are closely associated with centralized political power. Religious organizations can then establish a monopoly on specific ceremonies, or on the access to particular places like temples or shrines.

But there always is some competition. In all places with religions there is also a variety of alternative providers—personally identified specialists, such as shamans, healers, diviners, and mediums. In the imperial period, the official (state) Roman religion had to accommodate competition from the wild and exotic cult of the Great Mother, Cybele, whose devotees indulged in ecstatic rituals and self-flagellation. Their extravagant costumes were meant to contrast with the staid pomp of official ceremonial. Another source of competition was the dramatic spread of Mithraic mysteries and initiations.[20] Religious organizations can win many battles, mostly due to their access to political influence, but they seem to lose the war, as the resurgence of an alternative provision of religious services is inevitable.

This raises the question, Why do people resort to informal providers to supplement the services of official religions? The demand must be there, and it seems quite compelling, since people resort to such services even though they often are frowned upon, marginalized, or even prohibited by established churches or castes of priests. It seems that informal shamans, mediums, and healers respond better than religious organizations to a very specific demand that religious organizations do not meet.

One possible explanation, proposed by Harvey Whitehouse, lies in the contrast between doctrinal and imagistic practices. The doctrinal practice of religions is a form of intellectual training, accumulating a great number of relevant and explicitly connected propositions. It is characteristic of doctrinal religions. By contrast, the imagistic mode of transmission more characteristic of mediums and shamans consists of rare but exceptionally salient experience, so striking that its details remain engraved in memory. Doctrinal practices constantly run the risk of generating as much tedium as conceptual clarity, while imagistic ones can become so incoherent that most conceptual content is lost.[21] In his ethnographic study of Melanesian religious movements, Whitehouse showed that a shift toward a more doctrinal organization, with lessons, commandments, catechisms, and so forth, paved the way for imagistic revivals, with highly salient but conceptually ambiguous experiences. In the regions where Whitehouse did his fieldwork, people had adopted a new Melanesian religion, complete with specialists, coherent doctrine, rote learning of commandments, and codified ceremonies. Many

devotees then defected to join a splinter group that conducted highly salient, spectacular rituals with powerful imagery. This model would
suggest that religions and informal cults trigger different kinds of cognitive
responses. Religions provide coherence and explanation, cults memorable
experience.[22]

This contrast, however, may apply only to some situations of competition between institutional and informal religious activities. In many places,
people who consult shamans and diviners do not seek or receive imagistic
revelations from them. What they want and get are solutions to particular
problems, like illness, infertility, accidents, bad crops, and diseased herds.
In other words, the main benefit conferred by religious specialists may lie in
addressing specific cases of misfortune.

A universal concern in human societies is to explain the specifics of
misfortune, as opposed to the general laws that bring about untoward as
well as happy circumstances. The great anthropologist E. E. Evans-
Pritchard famously illustrated the point in his ethnography of the Zande,
taking the example of a hut that collapsed as people were sitting under its
roof. Most Zande people accept the general principle that explains the
collapse—they know that termites gnaw away the wooden pillars, which at
a certain point are bound to cave in. But they also want to explain why that
particular structure collapsed at that particular moment, injuring those particular people.[23] These are the questions a diviner is supposed to address,
and the explanation is usually couched in terms of witchcraft, describing
the accident as a deliberate attack on the part of specific individuals, who
wanted this to happen and made sure it hit the intended victims.

That may be one reason why religions generally fail to eradicate their
informal competition. Priests and their doctrines promote the notion of
large-scale superhuman agents, gods whose jurisdiction extends to an entire city-state, kingdom, or empire, affecting all and sundry in the same manner. Gods are described in terms of doctrinal principles that operate in very
generic terms. Sumerian and Egyptian gods are said to demand temples,
offerings, and ceremonies. According to the doctrines, the gods respond to
human action by protecting the city or the empire when they are placated,
flooding entire kingdoms or sending devastating plagues when they are not.

All this is about the society in its entirety, not about particular individuals. The ceremonies organized by official priests are supposed to guarantee the survival of the city-state or the empire's victory over its enemies, but they do not palliate the disaster of a particular farmer's bad crops. By contrast, the local superhuman agents that shamans and mediums claim to interact with are precisely involved in social interaction with particular people. These souls and spirits and ancestors are described as individuals with local connections, so to speak. The ancestors, clearly, are construed as concerned with what happens in their lineage. The souls and spirits of shamanistic ceremonies are said, for instance, to have stolen the soul of a particular person. Naturally, this contrast is something of an oversimplification. In many cases, the representatives of religions end up performing ceremonies for particular situations, and there is more of an overlap between the functions of priests and shamans than this model would suggest. Still, the divergence is real and may explain why there is always sustained demand for services besides those provided by established religions.[24]

Religions lose even when they seem to win. One great obstacle to understanding religious behavior is the belief, unfortunately widespread even among some social scientists, that people who officially follow a particular religion actually believe its doctrine. By contrast, one important discovery of cognitive studies of religious beliefs is what the psychologist Justin Barrett called "theological correctness." People claim they believe some officially approved doctrinal tenet, for example, that their god is everywhere and attends to everything at the same time. Then careful experiments reveal that, apart from this explicit statement, they intuitively expect their god to be limited in his perception, which of course makes sense since they construe their god in terms of their spontaneous intuitive psychology. So the official belief is stated, but it does no work, while a tacit, more intuitive belief guides people's expectations. This has been observed in very diverse religious traditions, from Christianity to Buddhism to Hinduism.[25] Theological correctness illustrates a principle that is generally true of human cognitive processes, though often ignored in discussions of religions, that people often do not believe what they believe they believe.

So, people may be given endless lessons on the fact that the Trinity includes three persons in one being, or that gods like mortals are part of an illusory world. They may even declare that they adhere to such statements. But the notion of a sentient person that is only one individual, has ordinary mental function, and combines that with counterintuitive physical features is much more compelling to human minds, because of its fit with our intuitive inference systems, combined with a salient violation of intuitive expectations. As a consequence, most Christians blithely ignore the unity of the trinity, and most Buddhists assume that the gods are quite real and should be propitiated. In this sense, religions also lose the doctrinal war, so to speak, in the sense that most people most of the time maintain theologically incorrect beliefs.

The Invention of Souls, Spirituality, and Salvation

What we commonly take to be necessary features of religion, like a doctrine and a priesthood, are recent developments that only appeared with the development of large-scale state societies with an extensive division of labor. Certain other aspects, which to many people may seem even more central to religion, are in fact even more recent inventions—that is the case with the idea that religious activities address spiritual concerns, like the cultivation of the soul, and that the soul, somehow, should be saved. Indeed, we know where these ideas originated, at what times and in what places they first emerged, before becoming very common in many of the world's organized religions.

Notions of souls and salvation are a hallmark of what the philosopher Karl Jaspers called the Axial Age, that period between 600 BCE and 100 CE when rather similar forms of religious doctrine appeared in China, India, and the Mediterranean.[26] These new movements emphasized cosmic justice, the notion that the world overall is fair, they described the gods themselves as interested in human morality, and these ideas came with all sorts of personal techniques or disciplines to do with moderation, self-discipline, and withdrawal from excessive greed and competitiveness. That is the case, despite obvious differences, with Buddhism, Jainism, and various forms of reformed Hinduism in northern India; of Taoism and Confucianism in China; and of Orphism, Second Temple Judaism, Christianity, and Stoicism in the Mediterranean.[27]

The cultivation of the self is perhaps the most intriguing aspect of these movements, which in very different cultures seemed to recommend very similar attitudes, notably moderate consumption, restraint from sexual excess, and the pursuit of a "good life" characterized by self-discipline and respect for others. The *Meditations* of Emperor Marcus Aurelius, inspired by the Stoic writings, provide a good example of that particular wisdom, which also echoes in the Analects of Confucius, most Buddhist texts, and many other writings of the time. These expressions of wisdom are very familiar to us, so it is worth noting how much they clashed with the prevailing aristocratic values in these different societies. For instance, the heroes of the *Iliad* are in this sense representative of the upper echelons of most ancient societies, in their pursuit of wealth, power, and glory, often in the cruelest ways. Against this acquisitive ethic, the new movements valued restraint and a withdrawal from worldly success.

The most important theme, which to this day shapes our understanding of religious activities, is the notion of the soul, as a highly individual component of the person that could be made better or purer and, crucially, could be "saved." The doctrines centered on the many ways one could eschew corruption or perdition of the soul. The prescribed way was to restrain oneself and lead a decent life according to established social norms (as in Confucianism), to demote pain and suffering as transient (as in Stoicism), or to renounce the material world altogether (as in Buddhism).[28] These movements created communities of ascetics or monks that would step outside society and cultivate the soul. In most cases it was understood that they would do that for the benefit of others, those unfortunate enough to be part of the material world.

To many people in modern societies, this view of the soul as the core of the person, in need of grace or redemption, would seem to be the core of religion. Even people who are otherwise indifferent to religious doctrines see the notion of the soul as crucial to "spiritual" life.[29] So the Axial Age matters, because the movements that appeared at that point in history had a considerable influence on subsequent religions. Indeed, the so-called world religions of today are all descendants of these movements.

So what explains the appearance of these doctrines, at roughly the same period in three different regions? A striking aspect of this development is that religious innovations appeared in the most prosperous societies of the time, and among the privileged classes in these societies. Gautama was a prince, Indian and then Chinese Buddhism spread primarily among the aristocracy, and Stoicism, too, was an aristocratic movement.[30] During the first millennium BCE, there was a considerable uptake in prosperity in the Ganges Valley in India, the Yellow River and Yangtze valleys in China, and the Eastern Mediterranean. Those places became much more prosperous than other ancient societies, such as, for instance, ancient Egypt or the Mesoamerican empires, as suggested by the multiple proxies for economic development measured by quantitative historians—for example, size of houses, amount of cereals produced, type of animal husbandry, size of granaries, size of towns and cities, proportion of craftsmen to farmers, production of luxury goods, and even pollution.[31]

But why would prosperity, and life among the upper classes of nobles and rich merchants, favor such ideologies? We have little more than speculative answers, whose value lies mostly in their parsimony and their congruence to independent evidence and accepted science. One possible explanation is a form of snobbery, whereby people signal their great wealth and status by ostentatiously renouncing (some of their) wealth and status, thereby signaling that they can afford such losses. As this is a common phenomenon in many distinct species, including humans, the cognitive machinery for such displays is obviously available.[32]

Another factor may have been that great affluence creates a situation of sharply diminished returns for some people acquiring more food or seeking greater social status and dominance. To those who have satisfied most evolved needs, an extra investment in such activities does not result in matching satisfaction. Individuals who have reached an extreme of relative affluence may become interested in doctrines that prescribe moderation and self-control, and feel the benefits of putting these recommendations into practice. People in such situations would spontaneously adopt attitudes of patience and long-term investment. As a result, they would find ideologies of moderation and preservation of the self intuitively appropriate and therefore compelling. But the explanation of course remains conjectural, given the fragmentary evidence.

Why did these movements expand to become most of the widespread religions of modern times? From the standpoint of modern believers, obviously, it is the truth and the spiritual value of the doctrines that explain their spread—but that is not historical scholarship. Another naive interpretation, based on hindsight, would be that these particular movements had some ingredients that made them culturally successful, that in other words they had found a set of beliefs that were most compelling, so that more and more people adopted them. That would be misguided, on two counts—because the religions based on these doctrines were generally not adopted but imposed, and because they were considerably distorted in the process, making them very similar to other archaic religions. First, the idea that people choose religions of course only describes a very recent modern situation, with no equivalent in history. In most societies at most times in history, people had no choice in religions, since the state, the king or the caste of priests, made that decision for them. The populace of course kept consulting diviners, healers, shamans, and suchlike, but those practitioners did not have the political clout to displace established religions. Second, when moral intellectual movements of the Axial Age became widespread religions, their contents changed. For instance, Buddhism spread first among the upper classes from India to China and Japan, and then developed a typical combination of ascetic ideal enacted by monks, combined with the charitable funding of temples and monasteries by the rich, and a much more pragmatic devotion among the populace, who generally stuck with anthropomorphic deities, amulets, and offerings. Early Christianity appealed both to the aristocracy, because of its Stoic flavor, and to the downtrodden, because of the eschatological message. But the most important factor in its diffusion was of course its adoption by Roman emperors and the considerable political influence of a well-organized church. Conflicts between these different groups, the aristocracy, the lower classes, and the ecclesiastical hierarchy, appeared again during the Reformation.[33]

Elusive Experience as Intuition

To understand modern forms of religious activity, we must consider another recent invention—the connection between religious beliefs and personal experience. In many modern movements, participants assume that religious activity should trigger a special kind of experience, entirely distinct from ordinary conscious activity, that these experiences carry important meaning, that they are crucial for a proper understanding of religious doctrines. Long before these recent developments, scholars in the study of religions, mostly in the West, for a long time argued that religious experience was quite special.[34] William James, the founder of modern psychology, also assumed that the nature of these exceptional experiences would be fundamental to understanding the emergence and development of doctrines and cults.[35] But it was not very easy, to say the least, to figure out what these experiences consisted of, and how they would connect to religious concepts.

As the scholar of religion Ann Taves argues, the comparative study of religions and modern cognitive psychology converge in suggesting that there is in fact no sui generis, specifically religious form of experience. However, all sorts of "special" mental events, which also occur in many nonreligious contexts, may provide potential anchors for beliefs about superhuman agency.[36] How does that happen? Very few specialists of religion have explored the precise process by which we could associate beliefs in superhuman agents with mental episodes that we experience as somehow different from the ordinary flow of conscious mental activity. A remarkable exception is the anthropologist Tanya Luhrmann's thorough study of a group of American evangelical Christians.[37] These evangelicals practice a specific version of mainstream Christianity, with a clearly articulated belief that God can talk to them.

But there's the rub—he does not. Or, to be more specific, the definite intuition that an agent is around, that this agent really is the god, that the god is talking, is a rare occurrence, and a frustratingly elusive one. Even among the most accomplished of believers a few islands of experience are surrounded by oceans of doubt and disbelief. From the outside, evangelicals are often perceived as people with certainties: they know there is a god, they know what he is like, they communicate with him. Inside the group, Tanya Luhrmann finds more or less the opposite. Christian beliefs are of course held with fervor, but the crucial elements, the presence of and communication from a superhuman agent, are described as goals to achieve rather than a starting point. Many evangelicals readily admit that they have not (or not yet) reached that point—it will take them more work.

The cognitive work takes many different forms, which constitute most of the community's religious practices. People must train their attention. In everyday life, we often experience transient, floating, and inconsequential intrusive thoughts whose origin is obscure and unimportant. For evangelicals, this is where superhuman communication may at some time occur, if one can train oneself to accept and ponder these elusive thoughts rather than discard them. People must also train their sensory imagination—their auditory imagination in particular, of course. They must seek places and situations where perceptions do not crowd out self-generated imagery. Openness to the god's presence requires careful monitoring and conscious appraisal of one's emotional experience. Most important, people must learn to pray. What most outsiders would consider the most straightforward activity, addressing an agent who you think is listening, is the most difficult, because the agent's presence is, precisely, highly problematic.

Practice works—somewhat, sometimes. Many members of the group studied by Luhrmann have experienced the "breakthrough" when inchoate thoughts or images seem to organize themselves into a coherent feeling of presence and a clear message from the imagined agent. Personality variables clearly help in the process, as Luhrmann's data demonstrate, but the main factor remains dedicated practice—one is led to the intuition of a god's presence through sustained effort.[38]

Why is this so difficult? The evangelicals described by Luhrmann are trying to put themselves in a particular mental state, in which they could literally hear a superhuman agent. Evangelicals also make their own lives and their faith very difficult, however, by spurning all the cheap tricks and devices that people the world over have used, for millennia, to induce altered states of consciousness. They do not want to open their minds to

the deity through the medium of drugs, starvation, meditation, hyperventi-
lation, or the hypnotic repetition of mantras. Which is of course why the
experience desired turns out to be so infrequent, ambiguous, and elusive.

These people provide a description of the sought-after experience
that is quite lucid and straightforward. This is exceptional. Most people
who seek religious experience, or comment on it, are much less specific
about the nature of the mental events concerned. But these inchoate experiences are still supposed to validate a specific doctrine or provide revelations
that cannot be achieved by other means.[39] People who value religious experience often contrast this direct form of religious activity with the unduly
intellectualized doctrines of familiar organized religions. For instance, in
modern Western contexts, people often assume that an emphasis on highly
special, private experience is characteristic of (vaguely defined) "Eastern"
doctrines and disciplines. (Ironically, some of the traditions that do emphasize private experience, like some forms of modernist Buddhism, were often
influenced in that regard by Western philosophy of religion.)[40]

The association between religious activities and special kinds of experience is very old, and so is the liberal use of what I described above as
tricks, those substances or practices likely to result in altered states of consciousness. In the context of traditional cults, without religious organizations, these special episodes could be seen as a means to receive direct
messages from spirits or ancestors, or to fight with them, as many shamans
do. These are all part of what Whitehouse called "imagistic" rituals, making
intensive use of exceptional, salient experience, with a strong suggestion
that the episodes contain crucial religious information. As Whitehouse
shows in his description of Melanesian cults, however, one cannot really
produce an exegesis of such episodes, which remain isolated memories of
exceptional experience with no clear connection to any precise contents.[41]

Could All This Be Adaptive?

Is "religion" adaptive? Obviously, no social scientist thinks that specific religious doctrines or notions are the direct outcome of evolution by natural
selection. But some have speculated that having one form or another of

what we commonly call religion might have positive effects on fitness, so that the propensity to adopt religious concepts or norms could have become part of our selected dispositions. Obviously, whether all this makes sound evolutionary sense depends a lot on what exactly we construe as "some form" of religion and what its alleged effects on fitness would be.

Here is one possible argument for the adaptive effects of specific religious notions. Some evolutionary anthropologists argue that aspects of religious behavior may constitute signals of commitment to a group. The argument is that religious activities seem costly if one considers, for instance, the time spent in prayer or ritual as well as the expense in terms of resources for rituals, sacrifices, and more generally religious pageantry, from the construction of temples to the upkeep of the priesthood. All this takes time and energy away from pursuits that would otherwise procure survival and reproductive benefits.[42] Precisely because of that cost, participation in such activities could signal that one is committed to the group. Here the anthropologists are being inspired by models of signaling developed in evolutionary biology. We know that signals can be honest, conveying an actual property of the organism, or dishonest, deceiving the recipient. For instance, a buck's large antlers honestly convey that the animal is strong; a cat's raised back hairs falsely suggest that the animal is large. Returning to human communication, commitment to a group could of course be conveyed by talk, but talk is cheap. By contrast, participation in costly, time-consuming activities like collective rituals might constitute more honest signals precisely because they are costly. People who engage in costly religious behaviors, for instance, submitting themselves to brutal initiation ordeals, giving away some of their wealth to monasteries or underwriting expensive ceremonies, would be signaling that they are willing to incur expenses for the benefit of their group.[43]

A problem with this interpretation is that religious activities may be less costly than it first appears. For instance, it may seem that sacrifice is a costly operation, and this is indeed the way it often is described by practitioners. People the world over commonly say that they will offer their best animals to the ancestors. But all this is of course metaphorical—the animals in question, once ritually slaughtered, are actually consumed by the participants in the ceremony. Moreover, these animals would have been slaughtered, ancestors or no ancestors, because they were bred for the purpose of being eaten. In other words, apparently costly religious behaviors should be evaluated considering the alternatives. If you did not sacrifice that bull to the ancestors (that is, actually consume its meat in the company of kith and kin, after asking for the ancestors' blessing), what would you do with it? The next best alternative would probably be to eat it without the blessing, and the next after that, to defer consumption to a later date; the worst option would be to eat it on one's own, outside a ceremony, thereby wasting most of the meat. So what is called a sacrifice (and understood as such in many places where it happens) actually is, in many cases, a not-too-costly economic decision, especially if we take into account the benefits one receives in terms of reputation and goodwill from other participants in the banquet. The same logic applies to many other examples of apparently costly religious activities. People who underwrite or sponsor large-scale ceremonies create debts and useful social connections, which is why this form of generosity is universal in human groups—those who have enough are motivated to engage in public demonstrations of this kind—whether gods are involved or not.

Still, there are times and places in which religious activities may be costly. People for instance build temples, spend considerable time studying abstruse doctrines, or give away their resources or property to religious organizations. That is not relevant, however, to our question concerning the evolution of religious propensities, because these costly behaviors are only found in the context of religions, that is, after the emergence of religious organizations with priests and doctrines—an event that only happened in large-scale societies, and therefore was much too recent (in evolutionary time) to have affected our mental design. There are virtually no such costly activities in the kinds of societies in which we evolved, no evidence for them from the anthropological or archaeological record.

But there might be another possible connection between religious beliefs and evolution. Religious concepts may contribute to people's fitness by persuading individuals to engage in generous cooperation with others rather than exploit them. In many places, ancestors and gods are described as powerful agents, who happen to be interested in people's adherence to social norms. They supposedly smite those who flout conventions, but also individuals who exploit others instead of benefiting the community. People who believe this may be inclined to be more generous than others, which would increase their fitness as they reap the benefits of cooperation and collective action. So, the reasoning goes, humans may have developed a tendency to find compelling the concept of monitoring by superhuman agents.[44]

Indeed, it is a common aspect of many religious traditions that people construe superhuman gods and spirits as "full-access" agents, that is, persons that have full knowledge of what people do, including what they do in relation to others.[45] So, is there an evolutionary advantage to such beliefs? It might seem plausible that social exchange could be more profitable if others saw their behavior as monitored by agents whose surveillance they cannot escape—this might rein in their propensity to cheat, if they truly believed that the ancestors' punishment would cost them more than they could gain by skullduggery. By the same token, if others could be confident that you do see yourself as monitored by moralistic ancestors, they would trust you, and you could reap the benefits of cooperation.

Talk, however, is cheap. You could always pretend that you do believe in the ancestors' surveillance, but in fact cheat without restraint, which would boost your own welfare, at least in the short term. The only protection against such hypocrisy is that signals of beliefs should be hard to fake, because they are costly. In the same way as buying an expensive engagement ring sends a more credible signal of romantic commitment than singing a serenade, a costly signal of belief in the ancestors would be better than mere words.

But then we are back to the previous problem—that religious behaviors in small-scale societies are typically not costly in that sense. People who participate in ceremonies for the ancestors seem to incur large costs only through sacrifice, which is not as costly as outsiders may think. People who consult shamans do compensate these specialists, but that carries no implication concerning their cooperative tendencies. Young men who undergo initiations do so as the only way to be considered fully adult men, a process that often does not involve any mention of ancestors or gods. So the anthropological evidence gives us no reason to think that people in small-scale societies could infer each other's commitment to specific religious beliefs, not to mention their cooperative character, from their participation in ceremonies.

What we know of religious activities in small-scale societies suggests that they contribute very little, if anything at all, to mutually advantageous cooperation. The anthropologists who argued for this connection were right that there are pitfalls to cooperation, that cheating is often more advantageous in the short term than cooperating, that there is an advantage in selecting partners with honest dispositions. But, as I describe in detail in chapter 5, humans managed to overcome these hurdles in ways that do not require that one ever evaluate anyone else's degree of belief in spirits and ancestors.

Perils of Functionalism

The notion that religion is there for something, that it is functional, is familiar to all—one thinks, for example, of Voltaire's contention that "if God did not exist, it would be necessary to invent him" (to keep the populace in check), a sentiment echoed in very similar terms by many other thinkers. In anthropology, the term "functionalism" described various theories founded on the assumption that social institutions emerge because they somehow keep a society cohesive. For instance, ancestor cults could be seen as practices that strengthened the authority of elders and thereby bolstered the social order of lineage societies.[46] One reason anthropologists abandoned this kind of explanation, notably as concerns religion, was that the hypotheses were essentially irrefutable. Another reason was that they failed to explain the actual contents of people norms. Any kind of cult, very different from the ones we observe, might strengthen the social order in some way—so why do we only observe these particular forms?[47]

The problem is much worse if we try to speculate about the evolutionary advantages of something as nebulous as "religion." The term is not really coherent, used as it is to denote supernatural fantasies, cults of ancestors, and interaction with spirits as well as spiritual experience and devotion to moralistic deities. But these different phenomena appeared in different places at different times, sometimes in combination and often not, so that there is no clear meaning to the question, What is the function of religion?

To illustrate how the question cannot be a scientific one, compare a very similar situation, that of sport. Does sport have an adaptive function? The question barely makes sense, because it is only by stretching categories to the point of vacuity that we could imagine that all human societies have some form of sport. True, in all human societies people like to play. In some times and places, people like to play and engage in vigorous exercise at the same time. In some places play takes the place of physical contests between individuals or small teams. In some places contests between adversaries attract spectators. Where does sport begin and end in all these various behaviors? No answer is better than any other, and the question is in fact not worth pursuing, as it is a matter of terminology, not of substantive understanding of what people do, in any of these different situations.

This does not mean that evolution is irrelevant to understanding human sporting activities, quite the opposite. Like other mammals, and in fact more than them, we enjoy play. That is probably because play activities help hone our musculature and coordination. Also, humans enjoy public displays of cunning and skill, probably because these advertise one's capacities and intelligence. Finally, humans eagerly engage in coalitional opposition, in forming teams opposed to other teams, as I described in chapter 1. These different capacities are all better understood in the context of human evolution.

The same is certainly true of the capacities and preferences engaged in the many different forms of religious activities. They all consist in evolved features of our minds. The capacity to represent counterfactual, supernatural situations is of constant use in our understanding of our environment. The ease with which we imagine nonexistent agents, or absent individuals, is certainly connected to our evolved social intelligence. The attention-grabbing power of some forms of ritualized behavior may be connected to the way we understand potential threats in our environments. All these will be better understood with progress in our understanding of the evolution of human cognition.

But that research requires that we leave aside incoherent terms like religion. As I said at the beginning of this chapter, anthropologists assumed that all human societies had religion, which they understood as a package that would include personal commitment, strong beliefs in metaphysical doctrines, a system of ceremonies, and a community of people with the same beliefs, as well as an explicit, coherent doctrine and organized special- ists. That was because anthropologists came from societies where such a package existed, and its existence was taken for granted. But in most other societies, and especially in the kinds of groups in which human minds evolved, there is no such package. As the anthropologist Maurice Bloch pointed out, the fact that religions are central to the institutions of many large-scale societies does not imply that it is special in cognitive or evolu- tionary terms.[48]

The Threefold Path

What is to become of religions? How can they survive science? Will reli- gions undermine civil society? It should now be clear how confused and confusing such questions are, mixing a whole variety of different social and cognitive processes under the term "religion," or in terms of abstract enti- ties, like science and reason versus religion, or religion versus civil society, or any other such conflict of abstractions. We now have better tools and a more precise understanding of the human cognitions and motivations in- volved. This does not entail that we can predict the evolution of religious activities, of course, but it does suggest several possible paths.

The first is the path of indifference. This is a situation in which most people evince no great interest in the doctrines or teachings of the different religions. Naturally, like other human beings, people in this context are still attracted to the products of supernatural imagination. Generally treated as fiction, these supernatural notions can sometimes lead to the "extraordi- nary popular delusions" discussed in the previous chapter, for instance, in beliefs about conspiracies or alien visitors.[49] But that does not imply any adherence to systematic religious doctrines, even less to membership in religious groups. Indifference to religions is not hostility. People in this situation have little motivation to attack religions, as long as religious organizations do not presume to interfere with their lives.

Indifference may seem puzzling to people for whom the existence and importance of religions appear to be inevitable features of human societies. But, far from being an oddity of some modern societies, the path of indifference is very similar to the attitude prevalent in the kinds of groups in which humans evolved. Religions appeared with large-scale kingdoms, literacy, and state institutions. Before them, people had pragmatic cults and ceremonies, the point of which was to address specific contingencies, misfortune in particular. They had little need for or interest in faith, cosmic gods, or a supernatural explanation for the origins of evil and the meaning of life. Most of human evolution took place in small foraging bands, where there is even less interest in religious matters as such, only an interest in the practical results that some religious activities can bring about.

Social scientists used to describe this indifference to religious doctrines and their prescriptions as typical of modern, prosperous European societies, but that was misleading. For instance, most Chinese people have little interest in religions in the Western sense of the term, as a more practical and ethical orientation has prevailed for centuries. America, too, may be following this path. Social scientists used to be intrigued by the "American exception" of a modern society where so many people are committed to religions, in contrast to the generalized indifference to religion that is characteristic of most of Europe. But the contrast was somewhat exaggerated to begin with, and in any case seems to be fading fast, as more and more Americans describe themselves as "unaffiliated" to any one religion. Remarkably, these people do not label themselves "atheists," which in America denotes a belligerent antireligious attitude.[50] They are simply not interested, joining the generations of Europeans and Chinese people who view different religions in the same way as most non-Canadians consider different hockey teams, as distinct groups engaged in an activity they have no interest in.

The second path is that of spirituality. The term is of course vague, which is rather apposite, as the beliefs people usually call spiritual are notoriously nebulous. Spiritual movements are focused not on particular statements about the world but on the exploration of various techniques and disciplines of the self. This spiritual orientation is, for instance, visible in the Western fascination for Buddhism, mostly in a very abstract version, focused on experience and mystical rumination, rather than on the folklore and practices of Buddhism in Buddhist places. The spiritual orientation comes in other forms, like the search for a sacred dimension of nature, Wiccan witchery, neopaganism, and the New Age potpourri of beliefs and ceremonies culled from tribal beliefs, shamanism, and various European mystical traditions.[51]

People engaged in such spiritual pursuits are generally not interested in the services offered by religious organizations. Indeed, many people who leave traditional religions see themselves as "spiritual."[52] They tend to think that most religions carry too much doctrinal baggage. People in modern societies know that factual statements from religions are embarrassingly misinformed—the world certainly is more than six thousand years old—and that the prescriptions are arbitrary—what could be wrong with eating meat on a Friday? That is why spiritual movements generally remove all the factual statements from religious traditions, replacing them with the vaguest statements about, for example, spiritual energies and planes of consciousness, which cannot possibly conflict with any modern knowledge. Also, the moral teachings of established religions are replaced with a focus on the individual's development, on the well-being or welfare of the soul or inner self. In this way, these spiritual fashions may be the true heirs of the Axial Age movements. Like them, they flourish in prosperous environments, mostly among people with greater certainty than most about being able to fulfill their basic needs. This spiritual turn motivates people to seek something not usually provided by the established organizations, but it is not clearly opposed to them. Indeed, spiritual activities are sometimes provided by marginal groups inside the established religious organizations.

The third path is the coalitional path. Affiliation to a particular doctrinal religion turns into ethnic or cultural identity and triggers the thoughts and motivations of coalitional psychology, including the clear separation between those who belong and the outsiders, the valuation of the group's collective goals, the assumption that the welfare of outsiders is a loss for the group, the close monitoring of other people's commitment, the attempts to deter defection by making it very costly, and so forth. Religious themes provide convenient themes for moral projects that center on the condemnation of a category of people, those, for instance, who eat "impure" foods or indulge in "unclean" practices. Relations between groups can take the form of moral crusades, focusing on the dangers posed by other groups or their activities.

It is tempting to interpret religious strife and violence in terms of ideas and extremism, to assume that conflict and violence occur because individuals hold extreme beliefs, such as religious doctrines that incite violence, or because the religious institutions will brook no discussion of their teachings, producing robotic minds prepared for assault against outsiders. What we know of human coalitional psychology would suggest the opposite explanation—that what people seek is coalitional strength and cohesion, and that religious themes that can favor moral recruitment will be intuitively selected when needed. The difficulties of the standard view, that extreme adherence to religious views is the cause of extreme behaviors, are obvious in cases where the doctrine cannot possibly justify the behaviors. That is to some extent the case in most religious organizations, and most salient in the example of Buddhist violence against non-Buddhists, in Thailand, Burma, and especially Sri Lanka. In Sri Lanka, Buddhist institutions gradually adopted a nationalistic and ethnocentrism rhetoric over the past century, initially as a response to colonial domination.[53] Numerous episodes of riots ensued, with monks and laypeople demanding restraints on Muslims and other minority religious groups, even asking for the demolition of mosques.[54] Obviously, this is especially striking given the emphasis on compassion and nonviolence in most Buddhist teachings, which were used, paradoxically, in ideological attempts to justify the violence as the defense of a pure religious identity.[55]

Coalitional dynamics explain the choice of themes to launch crusades, and also the competition that leads to extremism. In situations of high coalitional confrontation, people are motivated to monitor each other's commitment to the cause, but also to demonstrate that commitment to each other. Some individuals raise the stakes though conflict, because an extremely costly conflict, and a spectacular one, constitutes a convincing signal of commitment, and by the same token makes signaling ever more costly for other members of the coalition.

The coalitional path is also the most important one in terms of political consequences. A case in point is the change in relations between European Muslim communities and the host populations, as documented by the anthropologist John Bowen.[56] For some time, state authorities and Muslim groups in Britain and France tried, and largely succeeded, in creating institutions to represent specific Islamic interests in the political process.[57] At the same time, the relations between groups have become significantly more difficult. Many Muslims, with or without the support of their official representatives, have argued for a more visible presence of their values in the public space. That is generally seen as a shocking intrusion by host populations in both Britain and France, two places of massive religious indifference. In both places, Islam is now seen as intrinsically coalitional, which is why the vocabulary of "threat" and "contagion" becomes ever more common.[58] In a process of ethnification, that is, of considering categories of individuals as groups, with common goals and cohesive behavior, the situation is now widely perceived as a zero-sum competition for social control.

One should not take these three paths as an exhaustive description of the way religious representations could be handled by human minds. Nor should we think of the three paths as alternative and exclusive futures. They might coexist in the same place, and even in the same community. The difference between them lies in individual cognitive processes, whereby religious representations are mostly seen as possibly interesting fictions (indifference), as a way to cultivate the self (spirituality), or as the foundation of group solidarity and intergroup hostility (coalitions). We cannot, on cognitive grounds alone, predict the relative prevalence of these three paths. We can only be sure of very general probabilistic claims—for instance, that increased security favors indifference to religions, that some prosperity is required for spiritual interests, that coalitional recruitment is among the strongest forces in social interaction.

FOUR

What Is the Natural Family?

From Sex to Kinship to Dominance

IS THE NUCLEAR FAMILY THE NATURAL, fundamental social unit? Instead of trying to answer such a terribly misleading question, ponder this one: Should husbands live with their wives? Or this one: Do children belong to the same family as their father? In some places, the answer to both questions is "Of course not." Such dumbfounding exotica suggest how disparate cultural norms can be. But then, on a more sober note, anthropologists also tell us that in some respects human societies are very similar, that there is for instance (some amount of) male dominance in all known human groups, and that biological fathers everywhere have some connection to their children—so it would seem that there are common features to human families after all.

Such contrasted facts may feed intractable debates about the family, with some people arguing that specific norms of family life are "natural" and therefore in their view imperative, against those who see any mention of human nature as a snide attempt to legitimize very special norms. But these discussions are confused and confusing, relying as they do on notions of what is "natural" and "cultural." The best way to avoid this morass of disputation is to consider how the particular history of natural selection in the human line resulted in specific preferences and capacities. Starting from genes and sex allows anthropologists to ask important questions that used to be mysterious and intractable:

- What are the different forms of the family in human societies?
  Are there societies without families?

- Is there some form of marriage in all human groups? Why do these institutions exist?
- Why is there gender dominance? Is it universal? Is it always male? Why does it create terrible oppression in some places and times?

Do Other People Have "Families"?

Senufo husbands in West Africa do not reside with their wives. They visit them in the evening. They bring a few delicacies and spend some time with their spouse, before returning to their own family compound to enjoy the company of their siblings and other relatives. In other words, Senufo is a matrilocal society, where people are supposed to reside in their mother's group. It is also a matrilineal society, which means that children are members of their mother's but not of their father's group. Such arrangements are not the most frequent kind of kinship structure, but they are not rare either. Consider the Trobriand islanders described by the ancestor of modern anthropology, Bronislaw Malinowski. Triobrianders, too, recognize filiation on the mother's side. Children do spend some time with their fathers, in their early youth, but are then expected to move on to the uncle's group, their maternal lineage, first on the occasion of lineage ceremonies, and then as proper adult members of that group.[1] (A matrilineal society is not a matriarchy. The fact that descent is traced though mothers does not put them in power. In matrilineal societies as in other lineage societies, ultimate political authority is the prerogative of senior men. So, in political terms, it would be more appropriate to say that children become members of their mother's brothers' group.)

These cases highlight a simple problem. We commonly use the word "family," especially in modern Western societies, to talk about small units of residence, which typically include a couple and their offspring, what is often called the "nuclear family." We also sometimes talk of "extended families," which include grandparents or cousins. But if you consider places like Trobriand and Senufo, where is the family? Obviously, we cannot apply the Western notion of the so-called nuclear family to these places. It would not make much sense to say that somehow the father and mother and their off-spring constitute a group, a family. They have no common and exclusive property, nor do they reside together. They do not constitute a social unit in the eyes of others. The mother belongs to a particular clan, together with her maternal uncles, her mother and maternal grandmother, her maternal aunts, and their children. The same applies to the father, who is of the same group as his mother, mother's brothers, and so forth. There is no group that includes both husband and wife. Either nuclear or extended, the family is just not there. More generally, it is impossible to compare "families" across different cultures, because in many places the term is of no use. It would be difficult to locate where the family begins and ends. That is why anthropologists wisely chose to abandon the term "family" altogether.[2]

Indeed, a focus on the family may conceal a much more important and interesting fact about most human societies, namely, the paramount importance of kinship as a principle of organization. This is clearest in the fact that most tribal societies are composed of different clans or lineages that claim common ancestry. The most frequent type is the patrilineal system, where children belong to their father's and father's brothers' group, while their mother and her kin belong to another group. These patrilineal systems, often accompanied by patrilocal residence, are the most common form of social organization. There are other, more complex systems, like bilateral systems, where each individual belongs to two groups, traced through the two parental lines, and other variations on the descent principles. These people do not have "families"—but kinship organizes their existence. From foragers to small-scale horticultural societies, to agrarian civilizations and empires, humans lived in a social world largely organized around kinship ties. Different people's genealogical positions determined with whom they lived, with whom they shared or traded resources, who had power over whom, whom they could marry, and of course how possessions would be inherited. The pervasive nature of kinship is difficult to imagine for us denizens of modern mass societies, where genealogical ties are very short.

In traditional scholarship, the anthropological approach to questions of kinship was founded on an axiomatic separation between "social" aspects of family and alliance processes, on the one hand, and what were called the "biological" aspects of kinship.[3] This segregationist perspective, where facts about human evolution are relegated outside human cultures, is rather odd—and if we took it literally, we could not understand how marriage and filiation actually work.[4] Here are a few examples.

In matrilineal societies, there is often a tension between the claims of the lineage, on the one hand, and those of marriage and paternal filiation. That is, a man has political authority over his sister's children, and he counts her sons as his lineage members. But he may wish to offer help and support to his wife and his own children, even though they belong to another social group. The anthropologist Meyer Fortes described this kind of tension between contrary impulses among the Ashanti of Ghana, noting that people "are very much preoccupied by this problem and discuss it constantly."[5] As authority is vested in men, they manage the interests of a group that their children will never join, in the interests of their nephews and nieces. This tension is pervasive in matrilineal, and especially acute in matrilocal, groups. As a common result of this problem, marriage is generally less stable in matrilineal groups, compared to patrilineal ones.

This case illustrates a very important point, that many kinship and marriage systems are not harmonious systems of norms and concepts, in which every part makes sense in relation to the other parts. On the contrary, in many places the kinship organization is an unstable compromise between divergent motivations and norms.

Another example of such an unstable equilibrium is that of polyandrous groups, those rare communities where a woman may have several husbands. For instance, in high-altitude Tibetan valleys or in the Marquesas Islands, this practice was associated with intense cultivation of very small plots in a forbidding environment. Sons of a family jointly exploited the inherited plot of land, which remained a sustainable economic unit. So polyandrous marriage provided a solution to the problem of partition of inheritance—solved in other places by forcing all but one of the children out of the estate. It makes economic sense that brothers would stay together in places where agricultural expansion is not possible and male labor is in high demand.[6]

Polyandry, when it occurs, is not the consequence of increased power of some women, who would acquire many husbands. Rather, it is a situation where several men are so constrained in their choices that they accept to share a spouse. Anthropologists were for a long time intrigued by polyandry, because it seemed to go against an evolutionary imperative to go forth and multiply. As several men share the reproductive potential of a woman, the institution, it seems, would inevitably result in demographic decline. That was not actually the case, mostly because of a high rate of illegitimacy.[7] But even if polyandry is compatible with demographic expansion, it certainly comes with its share of problems. It relegates to an inferior status the many women who cannot find a group of husbands. Also, it creates conflicts about paternity. The norm usually is that all children should be considered the joint offspring of the group of fathers. But the reality is more complex, as people can often identify the actual fathers. In the Nyanja group, for instance, people indeed expect the actual father and their children to be especially close, and full siblings to be closer to each other than half siblings.[8] This shows that a kinship system may be full of tensions and contradictions. Indeed, no one in polyandrous places really seems to like the institution. When Tibetan people find jobs in the valleys and abandon their confined plateaus, they promptly abandon the practice.[9]

Again, these examples go against the old anthropological assumption that, in each society, there is a coherent set of cultural values or norms that make sense in relation to each other and provide people with representations of genealogical roles (mother, brother, sister, and so on) as well as their relations. Anthropologists and historians have long argued that the natural family did not exist. But they often replaced one myth with another, arguing that each human group or society had its own consistent model of the family or kinship system. That was just as misguided. In all human groups, the local kinship and family practices are a compromise between different preferences, inside each person and across individuals. For instance, a matrilocal system always includes some tension because men evolved to value investment in their own kin, which are more closely related to them than their sister's (actually, twice as closely). A patrilocal system, too, is a compromise, as a married woman needs to establish cooperative relations with nonkin, her husband's relatives, and forgo support from her own group.

But these examples also show that the traditional segregationist program, according to which people's concepts and motivations result from "their culture," and the culture in question cannot be connected to "biology," is just not coherent. Indeed, even though anthropologists often proclaimed this segregation, for instance, in textbooks and theories about culture, in actual practice the segregationist commandment was honored in the breach more than the observance. In their descriptions of the way kinship works on the ground, anthropologists would routinely enlist commonsense expectations, for example, that a father would like to contribute to his own children's welfare rather than to someone else's, or that a husband would rather have his wife all to himself rather than share her, even with brothers.

But there is a problem with this practice. The feelings and preferences that we find self-evident should, precisely, be seen as not obvious at all. Why would a father favor his own offspring over his sister's? Why would men be reluctant to share a wife? As I have mentioned several times, the great advantage of taking an evolutionary stance is that such familiar facts can be seen as odd, as something that requires an explanation. These facts of kinship can only be explained if we turn to the way human kin relations were shaped by natural selection.

Loops of Human Evolution

Over the past two million years or so, the human lineage has changed in ways that explain our modern ways of managing reproduction, parenting, and social groups. The story is a complicated one, because several processes of evolution took place at the same time, reinforcing or offsetting each other.

To disentangle these many causal links, let me start with the emergence of hunting. The gradual development of hunting had an enormous impact on human evolution, as it provided access to better nutrition, yielding not just more calories but also nutrition in the form of fat and protein, which are less abundant in plants. Access to a richer diet allowed the evolution of a larger brain with more complex cognitive capacities, so that brain size doubled between *Homo habilis* and the anatomically modern *H. sapiens*. Nutrition is crucial here, as the brain is the most energy hungry of our organs. But why would a more complex brain evolve? Among the many factors is the management of social relations. Having a more complex brain allowed early humans to track social relations and cooperation among many individuals, allowing for more efficient cooperation.[10] Also, larger brains allow more efficient hunting, especially of large mammals like deer, mammoths, or big cats, which requires that hunters acquire and store large amounts of information about the behavior of potential prey and develop cooperative tactics to compensate for their inferior strength. So hunting provided nutrients that helped grow the kinds of brains that are better at hunting. This is the first one of many feedback loops in this evolutionary process.

But a larger brain also means a larger head. Here natural selection hit a physical wall. Given the way the human pelvis is constructed, to allow us to walk on two legs, the birth canal imposes a limit on the size of the newborn's head. There are of course many possible engineering solutions to such a problem. What actually occurred is that early humans started giving birth to relatively premature infants, to deliver them before they were fully baked, so to speak, before the cranium grew too large.

This vastly increased the altricial character of the human lineage, that is, the fact that newborns are born helpless and require a long developmental process to reach maturity. Altricial offspring need intensive parental investment, particularly in the form of breast-feeding, so that, for a large part of their adult life, females would be either pregnant or nursing, which limited their participation in hunting. Beyond weaning, children would still require intensive care and protection, which again limited the females' capacity to secure resources. The availability of richer nutrition in the form of meat, however, would have compensated for that limitation.

Another feedback loop emerged with the invention of cooking, which breaks the cell walls in plants and attenuates their toxicity, as well as predigesting tough meat. These combined factors probably accelerated brain evolution, not just because of better nutrients, but also because of a reduced need for intensive digestion.[11] As the anthropologist Leslie Aiello suggested, brain and digestion were related in human evolution. The brain is an expensive organ—it appropriates about 20 percent of our energy intake, even though it is only about 2 percent of our weight. So early humans could increase their investment in that very expensive organ because they had a reduced need for equally expensive tissue in the digestive system. Indeed, a comparison of modern humans to other apes shows that we have a rather underdeveloped gut.[12]

Altriciality increased the cost of reproduction for human females. Part of the slack was taken up by others, in the form of cooperation in parenting. As the anthropologist Sarah Hrdy has argued, nurturing became to some extent a group affair, as many individuals, genetically related or not, would share babysitting and protection tasks.[13] A related and important evolutionary event was the appearance of menopause. In contrast to other primates, human females can survive their fertile lifetime by a long stretch, a feature that long puzzled evolutionary biologists. A longer life span and menopause created grandmothers, that is, individuals who could invest more of their time and energy in nurturing grandchildren, rather than having additional offspring. The evolutionary facts about the appearance of menopause are not as clear as we would wish, but this physiological novelty probably created another adaptive loop. Infants required more protection and investment, and mothers could provide it as they received extra help, which allowed for even more helpless infants and nurturing activity.

The Invention of Couples

A crucial evolutionary change was the emergence of pair-bonding, the close alliance between a man and a woman engaged in reproduction and parental investment. In all human societies there are such stable bonds between a man and a woman, with some expectations (however actually fulfilled) of sexual exclusivity, joint investment in offspring, and a large amount of unconditional cooperation and sharing of resources.[14] Even though all this is familiar to us, or precisely because it is familiar, we must keep in mind that these evolved behaviors are odd. True, pigeons, too, experience marital bliss—in fact many species of birds have stable reproductive pairs—but they are taxonomically very far from us. Among higher apes, the closest relatives of humankind, females are left to fend for themselves and provide for their infants, and that is true in very different reproductive systems—in the harems of gorillas as well as the promiscuous bands of chimpanzees.

Human couples are also exceptional in other respects. First, the bonds are often supported by strong feelings of commitment or affection between partners, as well as an intuitive sense of solidarity. Anthropologists have observed some form of romantic attachment and even passion in the most diverse societies—that kind of feeling is certainly not a Western invention.[15] The overlap between romance and marriage varies a lot between places and of course between marriages. Still, the sense of a common fate, of solidarity between partners, is remarkable and unique in our primate lineage.

Second, couples involve people besides the two main parties. The union of, say, Victoria and Albert creates social bonds between Victoria's relatives and Albert, as well as between his relatives and Victoria. In other words, human evolution invented not just couples but also in-laws. Indeed, in many human groups parents and relatives are actually involved in selecting the appropriate partner for stable unions—that is true of many forager groups, all agrarian societies, and many modern ones.[16] This would strike a chimpanzee anthropologist as extremely odd. In other species, there just are no in-laws.

Third, fathers are intensely interested in their offspring and are emotionally involved in their well-being and protection.[17] Fathers protect their offspring and direct resources toward them, but also play with them in many cultures and everywhere remain concerned with their welfare for many years. The birth of a child profoundly changes a father's motivations, a change that is reflected in neurophysiological and hormonal processes—fatherhood reorganizes a man's brain.[18]

These common features of couples make sense in the context of the evolutionary causal loops described above. Helpless infants required massive parental investment and made females less able than before to contribute to the acquisition of nutrients. In such a situation, females who could secure stable provision by a male would be in a better position than those who could not.

So a standard explanation of pair-bonds is that there emerged in humans a straightforward quid pro quo in which women offered (in principle exclusive) sexual access to men in return for sustained provision, in particular for those "expensive" foods that women could procure less efficiently than men, especially high-calorie meat from hunting.[19] This model, originally formulated as a "meat for sex" contract, attracted much criticism. Anthropologists pointed out that hunting in present-day foraging societies does not contribute much to people's diet. Also, in many foraging societies, strict egalitarian norms compel hunters to share their catch, so that it would apparently make little sense for a woman to expect favorable treatment from her mate. Also, what a child needs is a constant stream of nutrients, and what a hunter contributes are sporadic feasts. And large trophy hunting may be motivated more by prestige seeking than by efficiency in provisioning.[20]

These criticisms may be overstated, however. The poor returns on hunting are mostly true of modern hunter-gatherers, that is, people forced by agriculturalist pressure into the least productive environments. Also, sharing norms do not actually exclude favoritism. In many foraging groups, people will say that one should always share with all other members of the group, while actually being quite discriminating in their distribution. Finally, it is likely that meat was indeed an essential resource in ancestral environments. Even though meat may provide a small part of the calories needed, it delivers lipids and proteins, as well as many nutrients crucial to brain development.[21]

So there was a clear economic rationale for an ancestral division of labor, where individuals of each sex contributed more of what was comparatively advantageous to them. Women of course can (and sometimes do) hunt, but men are on average more productive hunters; men can (and often do) gather and process foods, but they are not more productive than women at the task. Economic reasoning would predict that, given these facts, some division of labor would advantage both sides.[22] Naturally, this did not require conscious deliberation. But pairs that divided labor in a more efficient manner would produce more and therefore achieve higher fitness.

It is true, however, that the "meat for sex" formula is a narrow and misleading description of this division of labor, for two reasons—because it's not just meat and it's not just sex. A crucial service that men provide in pair-bonds is protection against other men.[23] A female's fitness is always at risk from rape, abduction, and especially infanticide—the comparative evidence shows that in many species these are real threats for females.[24] These dangers are present in human groups, given male competition for access to females, particularly in the context of tribal warfare, which routinely involves (and indeed may be motivated by) the abduction of women from the enemy group. In modern contexts, too, an important source of danger for a woman is other men, and protection against them is assumed to be a component of the male contribution in a couple.

In return for resources and protection, a woman would provide . . . well, this is where the "meat (or anything else) for sex" formula is misleading, because sex is certainly not the one good provided in this quid pro quo, and this requires an explanation. Men's participation in stable pair-bonds evolved if it provided increased fitness. Any investment in children, from protection from enemies to babysitting to providing food, increases fitness because it makes it more likely that one's offspring will survive. Which is why there is so much paternal investment. But there is a crucial snag here, in the form of paternal uncertainty. Males cannot be certain that the offspring presented to them are actually theirs. A male who protects and provides for some other male's infants is acting against the transmission of his own genetic material. One expects that whatever genes result in such behavior would be selected out. Conversely, any genes that motivate males to be discriminating in their investment, protecting and helping those infants more likely to be theirs, would be at a selective advantage. Indeed, in many species one finds evidence of such male dispositions and capacities to gauge paternal certainty or to increase its probability. In those species, the motivation to provide for offspring is conditional on that certainty.[25] Humans are no exception in that respect.[26] So being part of a stable couple changes men's motivation, from simply seeking sexual opportunities to making sure one's partner is not seeking them elsewhere, which as we shall see may explain aspects of dominance between the sexes.

The Standard Model of Desire

Women, like the females of many frog species, tend to find partners with a deep bass voice more attractive than light tenors.[27] Men, like male goby fish, have a soft spot for partners with a nice shiny skin.[28] But attractiveness is more than looks and sounds, even among humans, and requires complex computations, of which we are of course blissfully unaware as we pursue the obscure object of desire. The loops of evolution that created helpless infants, helpful fathers, and stable couples explain a great deal of these computations.

In sexual preferences, as in other domains, nothing makes much sense except in the context of evolution, as Don Symons, a pioneer in the field, originally pointed out. From a fairly precise model of ancestral conditions, of what women required from men and vice versa, as well as of the division of labor, the energy requirements of nurturing viable infants, and the ecological conditions of the Pleistocene, evolutionary psychologists can generate hypotheses about sexual preferences that would have optimized fitness throughout human evolution. They can then test the models and sometimes reveal previously unknown regular features of human reproductive strategies, as well as explain already familiar ones.[29] Anthropologists and psychologists have accumulated a vast amount of evidence concerning sexual preferences and the way they interact to modulate partner selection, so vast that whole books would be required to do justice to the subject, and in fact several books do just that.[30]

Over more than three decades, psychologists and anthropologists have been carrying out a vast number of studies on the basis of these evolutionary hypotheses, in a large number of cultures and climates, with subjects of all hues and feathers. This large body of work has shown how fitness considerations make sense of many particular features of human preferences in the domain of selecting a mate, as well as predict some rather surprising ones. Mate choice is a rather subtle operation, one that requires a large set of complex algorithms.

Why does it have to be complicated? A first reason is that mate choice works at the margin. What matters for genetic fitness is not to attract a good mate but to attract the best possible one. As a result, attractiveness criteria will be skewed toward attending to small differences between individuals, creating a baseline of expected features, and paying special attention to deviations from this baseline. For instance, what usually makes a male voice attractive to many women is not that it is deep but that it is deeper than the average, which requires that some mental system compute the average. In the same way, men do not just like a partner with a smooth skin but prefer skin smoother than average, and the average is of course calibrated to local conditions. Criteria of physical attractiveness found in all human cultures, like a smooth skin in women or a square jaw in men, or facial symmetry in both sexes, happen to be very good cues of unobservable but crucial genetic and physiological qualities—but what matters to individuals are deviations from the average on these different dimensions.[31]

A second reason for sex being complicated is that the different mating criteria often vary independently. Since women evolved to expect both provision and protection from men, one would predict that they find attractive a whole set of distinct features. Social status, for instance, enters into women's criteria of male attractiveness but does not figure into men's computations of a woman's attractiveness.[32] As men are expected to provide for their offspring, important male attributes are ingenuity (which makes it more likely that a man will find resources) and generosity (he will share them). On the protection front, muscle mass (the potential for victory in fights) but also some aggressiveness (a will to use that potential) and selectivity in aggression should be attractive. Empirical studies show that these two sets of criteria are indeed used by women in assessing the mate value of potential partners.[33] But these two dimensions of attractiveness criteria (and there are many more) may not be strongly correlated, so that computations will include the relative weighing of different factors, which complicates the computational machinery required.

A third reason for complexity is that mate selection combines two kinds of preferences, derived from natural and sexual selection, respectively. In terms of natural selection, people should prefer partners that are most likely to beget healthy and fit offspring, which explains most of the preferences I mentioned so far. But there is also sexual selection, originating in the fact that the sexes do not incur similar costs in reproduction. In most mammal species the cost of reproduction is much greater to females than males. To stand a chance to reproduce just once, females incur the cost of gestation and the cost of nurturing an infant, which both redirect a great part of the individual's energy intake toward offspring. Also, as gestation and nurturing take time, no other reproductive activity can take place during that period. Male costs are much lower, consisting of mate acquisition, which may include some amount of violent competition with other males, the energy required to produce sperm, and of course the effort of having sex. As a consequence, females should be more picky than males, since the cost of mistakes is much greater for them. That is very much the case, including in humans, which is why, as David Buss, another pioneer in the field, once put it, sex is universally construed as something that men want and women may give.[34]

Female choice results in sexual selection, in the evolution of male features whose adaptive function is to respond to female criteria, to make it more likely for a male to be the chosen one. We are all familiar with the extravagant train of peacocks and the bright plumage of many other male birds—features that were selected because females preferred them to small trains and dull feathers.[35] These examples illustrate the surprising fact that sexual selection may go in a direction opposite to natural selection. Females may desire too much of a good thing. Peacock trains are heavy, and bright plumage makes camouflage difficult, so that exceptionally sexy individuals may not reap the benefits of attractiveness, having been exhausted by the effort or caught by predators. So males may develop behaviors that push right against the envelope of natural selection, for instance, by putting them in danger. Many human male behaviors conform to that prediction, consisting as they do of exhibitions of courage or resistance to pain that provide no direct benefit. For instance, bungee jumping started as land diving, a Melanesian display of male courage, in which men would climb up eighty-foot-tall towers and dive off, their ankles wrapped in vines that (if all went according to plan) would stop them just before they crashed into the ground. Sexual selection predicts that men would be motivated to engage in such displays, and women be sensitive to the qualities demonstrated, which the evidence supports. But sexually selected traits are not all about braggadocio. Sexual selection is influenced as well by the female preference for males who can and will provide protection but also nurture and protect their offspring, which is why males are motivated to demonstrate commitment as well as dominance or strength.[36]

Yet another source of complexity is that there may be different reproductive strategies operating within the same individual. We have long been a species of mostly serial monogamous pairs, together with exceptions that biologists named extra-pair couplings—called affairs in plain English. These two aspects, given our ecologies and our division of labor, correspond to two distinct routes to fitness, and two distinct sets of preferences. Long-term mating is what I described above as stable pair-bonding, the cooperative arrangement that combines economic solidarity, male provision, sexual exclusivity, and joint parental investment. Any features that make this arrangement possible and viable would be preferred. That is why women the world over are attracted to men with resources, and willingness to invest in offspring, on the condition that they can provide some signals of commitment. That is also why both men and women are especially attentive to the personality of potential partners. Most men intuitively know better than to reveal themselves as misers or cowards, which would destroy their value as potential mates. Women are intuitively aware that a history of promiscuity will make them less than altogether attractive to men in search of a long-term partner.

But humans also engage in short-term mating, which is very different, as it requires the partners to focus on what can be acquired here and now. In this context, women should prefer partners whose physique and status constitute proxies for "good genes," because of natural selection (having healthy offspring) and sexual selection (having sexy sons similar to their fathers, therefore with greater than average success at being chosen by females). That is also the context in which male preferences should be even stronger as regards cues of fertility—the sheen of youthful skin, the hourglass figure of nubile women. In men and women, these are precisely the shifts in preferences observed when individuals consider short-term mating.[37]

Finally, sexual preferences and attractiveness criteria differ across time and space, because the evolved systems are learning systems, sensitive to changes in the environment and making it possible to acquire information about these changes. For instance, men in many places seem to prefer women of a complexion that is slightly paler than the average—but this whiter shade of pale obviously cannot be the same in Iceland and the Congo. Accumulating body fat is a cue of good health, and therefore attractive, in challenging environments, less so in abundant ones. In the same way, the cues of dominance that may contribute to male attractiveness depend on the local political and social conditions. Powerful oratory plays in some societies the same role as humor, possession of a large number of pigs, or leadership in warfare in other locales. From the fact that sexual preferences should and do change from place to place, one should not infer that they vary randomly. Indeed, these variations result in choices that, on average, would have increased fitness in the environments in which we evolved.[38]

The scientific study of sexual preferences and behavior shows that humans are attentive to hundreds of distinct features. We need not be aware of them, which is all the better, as such complexity would overwhelm our conscious capacities. Our intuitions bear no trace of the immensely complex computations that triggered them. All we consciously experience, on most occasions, is that a face is singularly attractive or a personality unbearably winsome.

All this suggests that our commonsense view of sexual psychology, which also finds its way into much social science, is really misguided, when we think that sex is a matter of brutish instincts or urges of a very simple and direct nature. On the contrary, it is a matter of subtle calculations. In another oversimplification, social scientists used to think that matters of attractiveness and preferences were a matter of "sex" versus "gender" (one presumably more "biological" than the other). From the standpoint of evolutionary psychology, such a distinction appears laughably simplistic, perhaps even reductionistic to boot. Many computational systems are engaged in sexual preferences, identity, and behavior. They each focus on particular types of information and have their specific rules of computation. Cognitive scientists have barely begun to describe their interaction—what I described here was only a small part of the subtle and complex calculations that underpin sexual behavior.[39]

Why We Do Not Care for Fitness: Proxies

The pursuit of fitness explains many aspects of our sexual psychology, our criteria of attractiveness, and our motivations. But explanations in terms of fitness often seem abstract or counterintuitive, as no human being (or other organism for that matter) ever seeks fitness as such—to put it bluntly, no one cares about his or her own fitness. There is no fitness meter in the mind, a mechanism that would compute the effects of different behaviors on our capacity to produce viable offspring, and adjust our preferences to those consequences. To suppose that there is such a mechanism is a very frequent misunderstanding of an evolutionary approach to behavior, sexual behavior in particular.[40] But even commonsense observation should tell us that this is not really plausible. If our behavior was driven by a fitness meter, we would, for instance, be disgusted by contraception, and homosexuality would be unheard of.

The main reason there is no such mechanism in the mind is that fitness is largely invisible to an organism. Fitness is, roughly, a function of the relative frequency of one's genes in the future gene pool. But that is not something an organism can detect. Even if we took an easier but less precise proxy for fitness—how many offspring produced?—that would not help much. To evaluate the impact of their actions, humans would have to wait until their children could produce viable offspring. By that time, it would be rather late to adjust one's behavior.

So, rather than measure the elusive quantity of fitness, humans like all other organisms rely on proxies, that is, observable cues that would have been, on average, reliably associated with higher fitness in the environments in which they evolved. In contrast to fitness itself, these proxies are actual, observable features of the world. For instance, a square jaw is (to some extent) an observable indicator of relatively high testosterone concentration, which is (to some extent) associated with a willingness to acquire status and defend oneself and one's mate—desirable dispositions in ancestral and in many modern environments. That is why women tend to find male faces with that feature more attractive than others.[41] Note that the association between proxies and fitness need not be an ironclad certainty. A good probability is enough, such that individuals who were more disposed to find the feature attractive would, on average, be more likely to transmit their genes, including whatever genes influence that specific preference.

A good example of how proxies are used by mental systems is incest avoidance. In all human cultures, individuals evince disgust at the idea of sex with close relatives, and many official norms emphasize this rejection by describing all the horrific consequences of incest. In the 1920s, Eduard Westermarck argued that these norms were a way for humans to avoid the damaging consequences of sex with closely related individuals. Indeed, inbreeding has detrimental effects on fitness, mostly because of the risk of accumulating damaging recessive genes, and because it defeats the main effect of sex (and the probable reason why it evolved in living organisms), which is to shuffle genotypes—in order to present an ever-changing target to the myriad pathogens that evolve much faster than complex organisms.[42] Most species manage to avoid inbreeding, either through dispersion—moving far from relatives before being sexually mature—or by recognizing relatives by smell or other perceptual features. Humans do not have such direct perceptions of relatedness, and they do not disperse very far. Nor can men cast their seed to the wind, as many trees would do. Humans can, however, process much more information about their conspecifics than other animals—and that provides a solution. A series of studies by Debra Lieberman and her colleagues showed that specialized learning systems attend to information from co-residence, especially during childhood, and from one's own mother's interaction with the person considered. (There may be other cues, like physical resemblance, or even immune-system similarity.) This is used to compute a kinship index, a measure of relatedness, which in turn influences both sexual attraction (or rather the lack thereof) and a motivation for unconditional cooperation.[43] This explains the anthropological observation that unrelated individuals raised together are typically not attracted to each other. Their kinship inference systems are fooled by these exceptional conditions, and mistake the bride or groom for a sibling. That is why classical Taiwanese minor marriages, in which the bride grew up with her future groom, were less prolific and more likely to break up than standard unions.[44] In other words, behavior toward kin is influenced by a learning system that uses highly specific information in the environment to regulate our sexual motivation and our altruistic dispositions.

Our fitness is something that only some scientists could measure, some time after our demise. All that enters our minds is information about the environment, including other people, that can serve as cues to fitness. Each such cue, for example, the apparent intellectual skills, or body shape, or skin tone, or winning personality of a potential mate, triggers the operation of a specialized mental system, in a way that was selected because of its fitness effects.

How Environments Talk to Us: Life History

A great deal of our sexual psychology consists of learning systems that modify preferences as a consequence of acquiring specific information from the environment. This may also explain differences in lifestyle, not just between groups but also between individuals faced with different environments. The study of these effects is part of life-history theory, the field of biology that is concerned with trade-offs over a lifetime. At any point, organisms must allocate the limited energy available among such disparate functions as food acquisition, tissue growth, tissue repair, immune function, reproduction, and parental investment. Life-history models evaluate how these profiles are adjusted to optimize fitness.[45]

Budget allocations over time differ widely between classes of animals. Butterflies spend most of their lives as hungry caterpillars whose entire energy intake is directed to growth, before becoming butterflies that invest mostly in reproduction. Even within a class like mammals, there are salient differences between "fast" strategies, typical of organisms that reproduce quickly, have many offspring, invest very little in nurturing them, and "slow" strategy species, with longer lives, fewer offspring, and more nurturing. The typical profile for human life-history strategy is of course very much toward the slow strategy end of the spectrum. Humans require long periods of nurturing, their juvenile period is much longer than in comparable species, they have a long life, they invest energy for the long term.[46]

In recent years, biologists have pointed out that there are actually differences between individuals in terms of energy budgets, on the one hand, and systematic associations between specific environment cues and the adoption of specific strategies.[47] Some people seem to follow a faster than average strategy. They start having sex at a young age, have children early, have quite a few children, and seem more risk prone than the average person. Others delay sexual activity and childbearing, invest in their future, and seem more risk averse. These differences appear in physiological development, for example, the age of menarche, and in behavioral traits like impulsiveness. There is of course a continuum here, as most people are somewhere between the extremes of slow and fast lifestyles. These are not just a matter of conscious choices, obviously—young women cannot choose the time of their first period. The psychologist Dan Nettle was able to observe very large differences in life-history strategy, within a single city in England, that correlate with social status and affect people's behavior, notably their level of trust and future orientation.[48]

Although life-history strategy is partly heritable, individuals also modify their behavior as a response to environments, which can vary in severity but also in predictability.[49] The childhood environment is particularly crucial in calibrating individual life strategies, as harsh and unpredictable conditions orient individuals toward a faster strategy, with early sexual maturation and activity, early pregnancy for women, a more acquisitive and aggressive style that may result in antisocial or criminal behavior. By contrast, safe and stable environments seem to push people toward the slow end of that spectrum, with higher investment in the future, such as investment in education, and delayed reproduction.[50]

Obviously, the actual intricacies of reproductive and other life-history decisions are far more complex than this would suggest. In particular, it is perhaps too simplistic to think of environments along the unique dimension of harshness—different events, for example, famine as opposed to abuse or neglect, may impact our systems in very different ways.[51] War and famine lead women to suspend ovulation and therefore menstruation, as documented in history, for instance during the terrible Dutch famine of 1944–45.[52] In less tragic circumstances, excessive physical activity, like strenuous sports training, can have the same effect.[53] In both situations, a deficiency in fat tissue serves as an internal signal in the woman's organism that investment in reproduction is unlikely to increase fitness, as there is not enough energy available to support gestation and breast-feeding. The organism switches to an alternative strategy, temporarily favoring survival, in the form of immune function, tissue repair, and muscle mass to face adversity. This association between a specific cue (no father) and an inference about one's environment (paternal investment is unlikely) is a typical example of the kind of process whereby environments "talk" to cognitive systems designed to attend to highly specific information relevant to fitness.

Mysteries of Marriage

So far, I have considered how our evolved sexual psychology explains our preference and our dispositions, and how we respond to specific environments in predictable ways. But there is another aspect of sex and parenting that is crucial to human societies—the fact that humans in general stipulate norms of propriety, of sexual restraint, and of appropriate parenting. Why is that so?

The best place to start is the norm of marriage. The world over, people make a distinction between occasional or informal sexual encounters and arrangements (which may be approved, tolerated, frowned upon, prohibited, criminalized) and more stable and formalized unions.[54] The initiation of a formal union is generally marked by some public event. There are shared views about what each party should expect from the other, given such ceremonies, and about how they should behave toward third parties.

Sanctions are associated with the violation of these norms. From the point of view of an outsider to the species, several aspects of human marriage are rather mysterious.

Marriage is a package. Why do unions associate sex, children, economic solidarity, cohabitation? In other words, why would you expect to share food with people you have sex with? Why, after producing children, would you jointly raise your children and (generally) not other people's?

Marriage is a yes/no affair. Why a discrete step? The union that associates sex with cooperation and children is usually a matter of yes or no, rather than of degree. That is interesting because many other social relations are much less rigidly defined. One can evaluate them on a continuum—for instance, you could be more or less friends, more or less companions, but you either are or are not in a marriage. Even people who frown on traditional norms often end up creating equivalent notions of a stable, cooperative relationship, especially when they have children.

Marriage is for the long term. Marriages generally have no clear end point besides the death of the participants. This is not to say that all unions last forever, obviously. The union is generally construed as open-ended, as enduring for as long as no one does anything to terminate it. There have been a few historical examples of contractual, fixed-term marriage—but these are historical curiosities, whose rarity underscores how widespread, and apparently self-evident, open-ended commitment is. This is not the only domain where humans establish such open-ended relations (think of friendship), but here as in these other domains this feature should be explained.

Marriages require weddings. In most places, people organize some special events, often some ritualized ceremonies, to mark the inception of a union. True, in some forager groups people simply start living together and are gradually recognized as being some kind of unit. But in most human societies there are public events, and those are pretty conspicuous. Weddings are very audible and visible occasions—the norm being that the celebration should be as noisy and visually striking as possible. Why bother with all that expense and effort?

Traditional social science gave us a vast amount of evidence on vary-ing marriage practices, on which we can now draw to understand this very special human phenomenon. But we had no proper explanations. For in-stance, we were told that marriage was a rite of passage, something that marked the transition between stages in one's life, in this case from what could be called a social minor, still under the responsibility of some elders or one's group, to a full member of the group. But that just described the phenomenon. To take another example, many anthropologists argued that weddings are noisy and spectacular because the fact that people are joined in a stable union is a "social" matter, of interest to society beyond the two individuals within their families. That is certainly true, but it then raises the question of why the union between two people would be of interest to anyone else.

Even sophisticated social scientists used to take these aspects of marriage for granted. For instance, Gary Becker and other economists after him put forward a precise economic model of marriage, one where the costs and benefits for each party would be precisely described, from which one could derive predictions about actual social practices. The model pro-vided a remarkably clear descriptions of the conditions under which people (in a modern Western society) would marry, given the type of partner they would prefer, the number of children that would be optimal given their conditions, and so forth.[55] But this fine model also assumed precisely what we should try to explain, that people do want children, that they want to nurture them, that the children's survival is important to them, that they would prefer to share resources with their sexual partners rather than with strangers . . . in other words, all the features that I described above as mysteries of marriage.

Some aspects of marriage are not that strange, once placed in the context of the evolutionary loops that created hunting, cooking, helpless infants, and the sexual division of labor. Associating sex, economic solidar-ity, and the nurturing of children is a consequence of the evolution of highly cooperative pairs where sexual exclusivity and paternal provision are (in principle) assured, in such a way that they increase both partners' fitness. The existence of such pairs seems self-evident, as it fits with evolved templates for such cooperative pairs. By the same token, the evolutionary background explains why unions are of indefinite duration. Parental investment, in our conditions of evolution, required extensive cooperation between father and mother. But that cooperation could not have a specific termination point, a limited horizon, because the fitness of one's offspring is not clearly decided at a particular age.

But these explanations themselves raise another question. If humans form stable, cooperative pairs as a result of their evolved proclivities, why then do they bother to have marriage norms and organize weddings? What is the point of all this, if our evolved nature pushed us to form stable couples anyway?

One plausible answer is that these norms and interactions have advantageous effects for many participants, mostly in allowing people to coordinate their behaviors through communication. Consider the consequences of knowing that Victoria is now married to Albert. First, marriage conveys to third parties that the individuals concerned have withdrawn from the pool of potential mates. In other words, whoever had designs on either Albert or Victoria now knows that the time has come to look elsewhere. Second, marriage conveys to third parties that the individuals concerned have rights in each other that are not available to other members of the group. There is, for example, a certain amount of resources or help that Albert may expect from Victoria, and vice versa, or a woman from her in-laws but not from others. The fact of marriage reorganizes these expectations for all third parties. Third, marriage conveys to each partner that the other is (at least overtly) committed to fulfilling his or her obligations in accordance with the local norms. Fourth, it also communicates to third parties that they are so committed.

This would explain why people the world over expect marriages to start with public ceremonies, often as noisy and visually striking as possible. The pageantry has obvious communicative effects, as it conveys the identity of the partners, and the nature of their contract, to the largest possible number of outsiders. That is crucial because human pair-bonding does require commitment. Each party in a marriage may shirk the obligation to provide the expected goods or services. From a woman's viewpoint, reproduction is irreversible and requires investment in her offspring, but that is not the case for a man, who could desert after conceiving a child. Conversely, from a man's perspective, accepting a promise of sexual exclusivity is of course a bet on an unknowable future. The potential benefits of an efficient marriage in most cases cannot be achieved without sacrifices, as the spouses do not have identical preferences. So marriage requires honest, hard-to-fake signals of commitment. These are provided in many societies by costly conditions for marriage, for example, the obligation for brides to leave their kin groups, for grooms to provide bride wealth, to show adequate means to support a family, and the like.[56]

Making commitment public makes it stronger, because it makes defection more costly to one's reputation. Victoria cannot desert Albert and Albert cannot neglect Victoria without their breaking their word and revealing themselves as unreliable individuals, therefore unfit for cooperation in the eyes of third parties. That is a heavy price to pay—and it is sometimes paid, but it remains heavy and therefore makes defection less likely. This commitment effect of ceremonies may also explain why it seems natural to involve outsiders in the process. In the very simplified rites of Western societies, witnesses, best men, and maids of honor fulfill that function. In most other societies in history, a whole bevy of relatives would be involved. With more people as witnesses, the commitment effect is amplified.

Finally, these coordination effects between partners and between them and third parties explain why there is a specific category of marriage, a binary distinction between married and nonmarried, as opposed to a continuous spectrum of possible relations, all the way between occasional sex and full engagement in a stable cooperative pairing. There is a special label and a binary distinction because the coordination of behavior requires that different individuals know that they are coordinating, and commitment requires that they know they are committed. None of this, obviously, requires that people explicitly consider marriage institutions in these game-theoretic terms. People do not explicitly evaluate their institutions, nor do they deliberately plan and reform them. Instead, they happen to find some norms obvious, or legitimate.[57]

Gender and Dominance (I): Political Orders

Why are the men in charge? In most human societies for most of known periods, there existed a more or less marked asymmetry in power between women and men. This is manifest in different ways. Men have more influence on collective affairs than women. Also, in many societies, men seem able to control women's behavior to a greater extent than women control men. Finally, in some groups, the oppression of women takes a quite extreme turn, with restrictions on their freedom of movement, their control over their own lives—restrictions often enforced with utmost cruelty.

There is of course no shortage of answers to the question, Why gender dominance? Yet what matters here is not to review all those propositions but to examine the possible contribution of a naturalistic view of human behavior and capacities, taking into account what we know of human evolution and differences between the sexes.

Men have more power in the sense that they are more influential in, for instance, leading a village council or deciding when to perform the lineage's annual sacrifice to the ancestors. One may object, as many authors have done, that this is power but not official or overt or explicit power. That is, there is another sphere of influence that is not official or public yet has just as much impact on what actually happens in a group. That form of power relies on private connections, discrete influences, the establishment of networks of cooperation—and it is often handled by women as effectively as by men. That is the case, for instance, in many lineage societies where senior women are involved in negotiating marriages, and use them to build or strengthen alliances with other women.

But it remains that men more than women are in charge of overt, official decision making in most societies, including modern ones.[58] We cannot explain this in terms of authoritarian norms or patriarchal values, or of cultural norms. The problem is that such statements only beg the question of why people would adopt these particular norms or models—which is what we wanted to find out in the first place. It is more promising to consider the kind of psychology that could make dominance possible.

Men are in charge, but although the fact of male political dominance is very general, there is considerable historical and cultural variation in its manifestation. Two main factors contribute to this variation, the economy and the relations between groups—but the correlations are far from simple. One should obviously start with hunter-gatherers, as their foraging economy is the context in which we evolved. Even there, we find a great variety of situations. Foragers like the !Kung of southern Africa, who live in a rather poor environment, have little if any economic surplus and no clear political hierarchies. Women, like men, have a say in collective affairs.[59] This picture of a rather peaceful life, with fairly relaxed gender relations, is often taken as typical of our ancestral conditions. But it may well be an exception. In places with more abundant resources, like the Pacific Northwest, foragers had more complex political systems, with men in most political offices. The resources that made such groups relatively affluent came from men's work—fishing and trade—and men were in charge of relations with other groups. Among the Inuit, where men contributed almost all of the resources and there were raids between groups, women had very little influence on group affairs. In general, then, women's political influence seemed to depend on ecology and warfare, and the latter was an ever-present risk. Indeed, even the peaceful !Kung owe their relative peace to state domination, before which they had to respond to raids from the neighboring tribes.[60]

Women's political influence, varied as it was in foraging groups, was drastically reduced in agrarian societies, as subsistence depended on heavy work mostly provided by men. Indeed, women's status seems to be lowest where agriculture is based on using the plow, which requires male upper-body strength.[61] A sharp distinction between men's and women's work is typical of agrarian societies, where men contribute the bulk of subsistence and are in charge of the fields, of large animals, and of relations with other groups, while women manage the domestic sphere.[62]

Through all these historical changes, one straightforward prediction about human societies is that if there is a sexual imbalance in political influence, it is in the favor of men. As this is found in the most diverse economic and ecological environments, the difference predates historical developments like the appearance of agriculture or large cities.

One important factor here certainly is the clear division of labor that emerged during our evolution. Reproduction and parenting involved stable pairs in which the male provides crucial complements to gathered foods, but also protection against other males and against other groups. As I mentioned in a previous chapter, we should avoid the symmetrical pitfalls of Hobbes's vision (a war of all against all) and Rousseau's (cooperation between peaceful Noble Savages) in our descriptions of ancestral conditions. More soberly, the evidence suggests intensive cooperation within groups and potential conflicts, including warfare, between them. If that is the case, it would follow that men's decisions were the most crucial ones for social groups, like bands or tribes, as groups rather than as collections of individuals. In other words, the imbalance in terms of politics would be favored by psychological differences, which themselves can be traced to the fact that politics, for most of our evolutionary history, often came down to the single question of whether to go to war with neighboring groups or whether to expect an attack from them, and to evaluate whether they could be pacified, notably through trade.

The reality and importance of primitive warfare during our evolution suggest that some aspects of male psychology would be adapted for intergroup conflict. Several kinds of evidence support that prediction. Differences in aggressiveness between men and women, as well as the upper-body strength required for combat, suggest selection for intergroup violence as well as competition with other men for access to women. Also, some sexual differences suggest that male minds were shaped by intergroup rivalry. For instance, in economic games where people can contribute to a common pool (so-called public good games), men contribute more in the context of competition between groups than in competition between individuals— while this makes no difference for women.[63] Male and female minds construe cooperation in different ways and recruit different brain circuits to manage it.[64] More generally, women tend to construe social relations primarily as between persons, while men readily view them as between groups. Men and women may even recall the same events differently, from a group and a personal standpoint, respectively.[65] These differences appear early in childhood, as girls and boys in the same school environments create different kinds of networks—with fewer, more deeply committed links among girls, and more numerous but less stable recruitment among boys.[66] The differences persist in adults, even in business environments, as men and women create different kinds of networks.[67] Sexual differences in capacities and motivation, then, would confirm that one crucial fact of human evolution was the role of men as warriors, and by extension as managers of relations between groups.[68]

To go further, and this is speculative, our ancestral conditions may also explain the difference between the overt, official, often ritually expressed politics of men and the informal influence that women wield in so many societies. Men and women both need stable alliances with friends and supporters, but they needed them in different ways and for different purposes in our evolutionary past. Women needed to recruit allies for collaboration in food extraction, and for help with parenting. These activities require few people, but with deep enough commitment for long-term cooperation. By contrast, men's collective actions included hunting and warfare, two activities in which one needs to mobilize larger groups of people—certainly more than two or three individuals. Hunting and particularly warfare, especially "tribal" warfare, require excellent coordination between parties. That is, everyone must be aware of what the others should be doing, and be able to monitor whether they are actually doing it. Finally, both are dangerous activities where defection is very costly, as one's survival may depend on others taking risks as promised. This means that other individuals' commitment to the group's enterprise must be gauged and carefully monitored.

Different ways of cooperating may require different kinds of information flow. To maintain small-scale friendly networks, one needs access to individuals as such and one needs a measure of discretion. Every item of information need not and in many cases should not be broadcast too widely. But large-scale coalitions for hunting or group defense often require overt, public announcements so that all participants can better coordinate their behaviors. Just as important, publicity serves as a guarantee of commitment. To the extent that people pledge allegiance to a cause in front of many interested parties, they incur a large cost in terms of reputation if they ever defect. Parading with the rest of the militia can serve as a commitment signal, a costly signal that one will stick with the other fighters, even before any engagement has taken place.

This of course does not entail that the social worlds of men and women are entirely exclusive and different. Indeed, even in small-scale economies, when people clear land for gardens or cooperate in butchering a large prey, they typically do so in large, all-inclusive teams in which male and female participation may be similar. In many situations, in fact in most situations in modern societies, men and women need a support network that includes both genders. But the difference remains: the evolutionary pressure toward small-scale cooperation through bilateral ties was probably stronger on women than men, while pressure for larger-scale, multilateral group-level cooperation was conversely stronger on men—although we do not yet know, for lack of systematic studies, to what extent these differences are related to the fact that the two genders participate in group politics to a different extent, and often in a different manner.

Gender and Dominance (II): Domestic Oppression

Domestic oppression consists in limitations on the freedom of movement of women, their choice of dress or sexual partners, and many other constraints on their autonomy. Here, again, it seems that our traditional explanations cannot get us very far—simply describing this form of oppression as the consequence of stereotypes or even hatred of women. These descriptions are not explanations.

Restrictions on women's autonomy take many forms, including the well-known purdah or hijab, those literal and metaphorical curtains behind which a woman should remain hidden, confined in the home, obligated to conceal hair or skin or eyes or the entire body. Constraints of this kind are found in many agrarian societies, but they are pushed to an extreme in the modern Middle East and the Islamic world—including, for instance, a ban on women driving cars, the obligation for a woman to be accompanied by a male companion at all times in a public space, or the requirement that a woman obtain a male relative's approval for her to seek employment, get identity documents, or travel outside her neighborhood.[69] A particular focus is women's dress, with official or informal norms about the extent of skin that may be shown, and about the propriety of showing one's eyes or hair.

Despite their variety, these constraints all restrict as far as possible a woman's access to men other than her immediate male kin and her husband. The fact that such access may lead to sex is a leitmotif in discourse about women in the societies where such constraints are the norm. We cannot explain these oppressive measures as simply an attempt by men to exert power for the sake of dominance. That would not be plausible, as we would then need to explain why the restrictions, by extraordinary coincidence, invariably limit a woman's sexual autonomy, even though they vary greatly in other domains. There is no place where women are constrained, for example, in their movements in the public place or in their form of dress, but on the other hand have sexual autonomy. In all places where there are limitations on women's personal autonomy, their sexual choices are restricted—there is no exception.

Perhaps it makes sense to see this in the context of evolved mate-guarding behaviors. In many different species, males invest considerable time and energy to make sure that their female mate has limited access to male competitors, for example, by following her around and threatening competitors. Male baboons, for instance, forgo occasions to get food in order to stay close to a female they are guarding. Male warblers spend much energy monitoring their female partner instead of foraging.[70] Two factors determine the intensity of mate guarding. One, obviously, is intrasexual competition. Mate guarding by males is more intensive in species where there is a large asymmetry in reproductive capacities—that is, where a few powerful males monopolize access to females. Male mandrills, for instance, are very different from females, larger and more colorful, which suggests intense intrasexual competition. Males are exceptionally competitive, such that a dominant male may be involved in more than two-thirds of copulations in the band, but at the cost of constant monitoring of females.[71] The other factor relevant to mate guarding should be paternal investment. To the extent that a male invests in protecting and nurturing offspring, cuckoldry is a major fitness threat. That is the case in many species of birds, which is why males invest time and energy monitoring the females.

The same principles apply to humans. We can infer from indirect cues, like the difference in size and strength between males and females, that humans had a moderate but real amount of sexual competition. More important, long-term, intensive paternal investment is a characteristic of human pairs. This would predict a great amount of mate guarding in humans, as certainty of paternity is all the more crucial to male fitness—and that is indeed the case. David Buss and other evolutionary psychologists have documented the many mate-guarding techniques used by men, like concealing the existence of a partner from friends and acquaintances, monopolizing her time, and denigrating possible competitors, threatening or assaulting them.[72] From fitness considerations, we should then expect that men will monitor women all the more closely when they are of fertile age or are perceived as greatly attractive—all predictions that are supported by experiments and observation.[73]

Sadly, another confirmed prediction of a fitness interpretation of mate guarding is that it may lead to violence against the partner herself. Obviously, many distinct factors lead to violence against wives or partners. But the careful studies conducted by Margo Wilson, Martin Daly, and others after them clearly show how evolved motivations drive these behaviors. Violence is a deterrent. That is, it is not simply a result of men being more violent than women in general, or of an urge to exert power in the domestic domain—crime statistics show that violence toward a female partner is more likely when the victim is younger or more attractive, when the presence of stepchildren provides a cue that she may desert the man, and when the man himself is of low mate value, making desertion more likely.[74] In other words, violence occurs more where the perceived need for mate guarding is higher. Wilson and Daly gave the term "male proprietariness" to this complex of evolved motivations that translates access to a woman into an illusion of exclusive ownership and triggers a whole suite of behaviors designed to deter defection and poaching.[75]

Mate guarding is supported by a learning system that attends to specific cues in the environment, for example, about the presence of other men or the partner's interest in them, and combines them with information like details of one's own mate value, one's partner's, the state of the mating pool, and so on, to modulate behavior, between relaxed confidence and anxious monitoring, or even violence. That is why we may expect mate guarding to take on very different forms, not just between individuals, but also with large differences in ecology and social interaction. For instance, in places where there is a great economic inequality, we could expect that men lower in social status also have more to fear, and therefore would be more motivated to restrict their partner's freedom, as observed above. Or consider the differences between foraging and agriculture. Many foragers live in small bands, in almost constant contact with each other, which makes it easy for people to monitor their partners—although not infallibly, of course. But an agrarian society provides a very different social environment. Men and women work apart from each other, and in many cases men work alone— which provides no information on other men's whereabouts and activity. Urban environments provide yet another kind of social ecology, with multiple unknown individuals in potential contact with each other. So we should expect that the constraints on women's autonomy that result from mate guarding also vary with these ecological differences. There is unfortunately no systematic study yet of these differences, and of their interaction with individual variables—so that this part of the explanation of domestic oppression is still speculative.

Socialized Oppression as Collective Action?

There is still an important question left unsolved, however. Consider this example. A few years ago, a young girl was assaulted in the Orthodox Jewish community of Beit Shemesh near Jerusalem. Being from an Orthodox family, the girl was dressed in what most people in Israel and the rest of the world would judge to be an extremely modest fashion. Apparently, that was not enough for a group of enraged young men, who surrounded her, spat in her face, and called her a whore and other names. The main source of their righteous anger was her bare arms. She was eight years old.[76]

The incident became a journalistic sensation, mostly because it happened in Israel, a largely secular society, where the extremism of fundamentalists is a perennial concern and an irritant to many citizens. Thousands of scandalized people joined demonstrations in several cities to denounce this eruption of puritanical folly. In many other countries, such incidents are frequent and generally go unreported, and very often unpunished. In many places in the Middle East men routinely gang up on women who fail to dress according to their standard of modesty. Women can be harassed or assaulted for violation of some regulation on what they should wear, say, or do.[77] There is popular approval in many Muslim countries for constraints on women's autonomy, with support from state institutions, although that of course does not extend to condoning actual assault.[78] Still, harassment and attacks do occur, despite informal protests and the more organized resistance of some women's organizations.

There have been many anthropological debates on the history and dynamics of women's rights in the Muslim world, challenging a simplistic description of the local values as uniquely repressive and of the women as just passive victims.[79] I shall not review these debates here, as I focus on a much narrower phenomenon—the fact that men, in a public space, can claim a right to police the behavior of women and threaten rebellious ones with violence. The phenomenon is certainly not unique to this region of the world. It would seem to challenge our explanation of women's constraints as a form of mate guarding.

Going back to the incident in Beit Shemesh, an unsolved question is, Why did these men bother? Not one of the fundamentalists who assaulted the young girl was her husband (obviously). But then, why would they feel concerned for her modesty? The policing of women's behavior by perfect strangers is a puzzle. To see why it is odd, consider its effects in terms of fitness. Although males in many species engage in mate guarding, they do not guard other males' mates. A man who participates in reducing women's opportunities may protect another man from cuckoldry, and therefore increase that unrelated male's fitness. It is unlikely that natural selection would favor such dispositions. That is why the behavior is puzzling. And that is not the only puzzle. Another is why this happens in public, and in a spectacular way. It is not just that the men engaged in such policing want a woman to behave in particular ways—they want her to be shamed for behaving differently, and they are happy to be seen doing it. To add to the mystery, reports from these incidents suggest intense anger directed at the women, which translates all too easily into violence. Why?

Perhaps this can be explained in quasi-economic terms. For each man's fitness, certainty of paternity is a crucial goal. In many situations it is very difficult to achieve. In theory, the only situation that could guarantee it would be the partner's complete seclusion, under constant scrutiny. Indeed, throughout history powerful men who could afford to do so locked up their wives in closely guarded harems.[80] But that requires exceptional resources. Most husbands in agrarian societies or urban conditions could not bear the cost in time, energy, and personnel. Also, women in villages must work in fields or gardens, those in cities must procure food from the markets. And men are often engaged in economic activities that separate them from their partner. So it is clear that a husband cannot carry out all the desired monitoring.

A man could, however, trust that other men will do it for him, on condition of course that he reciprocates. If such cooperation is possible, it may bring the benefits of surveillance at a much-reduced cost. Each man participates in the general monitoring of women, in enforcing all the petty rules that confine them to specific places, modes of dress, and so forth. That is not, in most circumstances, a very costly investment, because it is distributed over all the adult males in the local community and because the victims of the system rightly fear punishment for transgression, and therefore do not usually test the tolerance of the men. As the investment is not really costly, one would predict that men would engage in it, inasmuch as it offers some potential benefit. The benefit, here, is the assurance that one's own actual or future mate is also being watched in that same way. In other words, a husband does not need to monitor his wife at all times and in all public places, because he can be assured that, at all times and in all places, there will be other men to do the job. Most husbands benefit from the system, as it reduces the costs of surveillance. Most husbands contribute to it, by accepting to monitor the behavior of women who are not their partners.

If this explanation is valid, the tacit pact amounts to socializing oppression, turning it into a form of collective action. Each should contribute, and all will receive. Obviously, this collective-action interpretation does not require that anyone be aware of the pact and its conditions. As in other forms of collective action, like collective hunting or warfare, people do not explicitly go through the game-theoretic computations of their costs and benefits, because mental systems do it very efficiently, away from conscious inspection. All one need be aware of are the motivations, for example, to help others during the hunt, to accept risk, and so forth, and in this case to monitor women and shame the rebels.

We know, however, that collective action is vulnerable to free riding. For instance, it may be advantageous for all workers to go on strike and obtain higher wages. But for an individual worker, it is even more advantageous to avoid striking and to reap the collective benefits. So collective action will unravel unless specific conditions are met. And we should expect incentives for free riding when the collective action consists in organizing women's oppression. Whenever a man encounters a woman who, given the local norms, seems to signal sexual availability, for example, by the extreme expedient of showing a lock of hair or a forearm, he might try to take advantage of (what he takes to be) her disposition, rather than enforce the norm for another man's benefit. If that is possible, there is no incentive for any man to participate—he would be helping individuals who will not help in return—and cooperation will soon unravel. So it takes very special conditions for this equilibrium of cooperative mate guarding to persist. But in some societies, it seems that the conditions are present.

For one thing, note that the constraints described here are imposed on women's public behavior. They must not wear the wrong garments or talk to unrelated men in the public space, where all behavior is observed by strangers. As this public space is the only place where a man could encounter an unrelated woman, it is difficult for a man to engage in the opportunistic behavior I described above. Were he tempted to do so, he would have to break the norm of male surveillance in full view of the public—in other words, to betray the collective action in a way that cannot be plausibly denied. Also, the fact that this all happens in public means that each man who scolds a woman does it in a way that adds to his reputation. That of course is crucial in any collective action—each member needs to persuade others that he is committed to the collective goal, for fear of being seen as a defector. By participating in the public rebuke of a girl who shows her bare arms, men signal to each other that they are truly committed to the collective goal.

Another factor that helps socialized oppression is that, in some places, the norms of propriety are very clear, and known to all. Everyone knows what precise amount of flesh can be seen, what colors are allowed, and so forth, even though women are, naturally, constantly pushing against the limits of petty oppression. As the norms are clear, it is easy for any man to determine that a woman is transgressing the rules, in which case something ought to be done. This also makes it much easier for each man to know whether another man defected from the collective action by tolerating a particular transgression.

Last but certainly not least, in some countries men who engage in collectivized oppression rightly assume that state institutions, like the police and the judiciary, will be on their side and are very unlikely to defend the women, even in cases of harassment or violence.[81] In economic terms, this decreases the cost of participation even further and therefore makes it more likely that people will participate.

This is, to a large part, a speculative model, as there is little experimental research on the psychology of oppression in these contexts. But the interpretation in terms of collective action seems to make sense of some otherwise puzzling features of collective monitoring. For one thing, it would of course explain why there is an incentive for many men to participate in boosting the fitness of other, unrelated men. Explanations in terms of shared values do not solve that puzzle. When we say that men monitor women because they want to impose patriarchal models of chastity and modesty, we simply assume what we had to explain, that a man would be motivated to impose these models on women for the benefit of other men. Also, the collective action model explains why men who participate would be motivated to do it in public and to shame women rather than simply get them to change behaviors. Why the public fracas? This makes sense if men feel the need to demonstrate their participation in collective action. The loud and public demonstration may be largely for the benefit of other men, as a demonstration of commitment.

In the same way, the collective action model may help make sense of some participants' anger. We must, again, remember that the computations that lead to specific emotions are not conscious but do nevertheless consist in calculations of costs and benefits. Anger is usually triggered by an intuition that we are being exploited by others—in other terms, that they are not valuing our welfare as we think we could expect. That is, anger is not an outburst of irrational, undirected energy—quite the opposite. It occurs when we (or rather, our mental systems) detect exploitation but also calculate that it could be redressed by the threat of retaliation.[82] This may well be why at least some men are angered by women who dare to defy the rules, and thereby send a signal to others that it is possible to do so. For a man who participates in the collective monitoring of women, the presence of a rebel in public space shows that the system does not work, that his own investment in it will not guarantee his benefit in the form of certainty of paternity, in other words, that he is being exploited—which may account for the rage at such disobedience.

On a less disheartening note, this collective oppression model would explain why such systematic monitoring and harassment of women is bound to be rather rare. In most societies in the world, we do not observe the implacable policing described here. That is probably because the conditions for such efficient collective action are missing. True, many men, as husbands, would probably benefit from such practices, in the form of increased certainty of paternity. They would benefit even more, however, if such norms were upheld by others while they took advantage of transgressions. Together with women's resistance, this standard obstacle to collective action, the potential for free riding, is fortunately there to limit the spread of socialized oppression.

How Can Societies Be Just?

How Cooperative Minds Create Fairness and Trade, and the Apparent Conflict between Them

TO CITIZENS OF MODERN, LARGE-SCALE societies, these are probably the crucial issues of politics: How can social and economic systems provide justice? Why is there inequality at all? How much of that inequality is morally just? An early version of this modern preoccupation is Jean-Jacques Rousseau's *Discourse on the Origins and Foundations of Inequality among Men,* which pointed to the very existence of private property as creating the conditions for inequality.[1] Our modern concerns about justice and society, very much like Rousseau's, are rooted in our understanding of the economy. The question, What is a just society? is clearly construed as a question about who produces what goods, who has access to goods, under what conditions, or to what extent, and how the rules under which we interact with others may create fair or unfair differences.

And this is a question for human evolution, because natural selection explains many aspects of what we understand by justice in society. First, it explains why we have a sense of fairness, why it manifests itself in similar ways in different human minds, and why it triggers such intense emotions. Second, it also tells us why humans cooperate, exchange, and trade, and what capacities make it possible to create gigantic interactive systems like modern economies. I know that both claims may seem unintuitive, to say the least. Aren't moral norms something that we get from society, from being raised in a particular cultural environment? Don't they differ a lot from place to place? As for extensive trade and markets, they are obviously very recent, in the time scale of genetic evolution. How could natural selection explain how they work? But the evidence suggests that human evolved psychology does in fact provide us with a way of understanding both our concern for justice and the emergence of mass-market societies.

Cooperation as a Mystery

Humans are immensely cooperative. Precisely because we are so coopera-tive, and because cooperation is part of our evolved nature, it is often diffi-cult to see it, and to understand how it organizes our behavior. To emphasize the point, Sarah Hrdy describes how passengers board or leave an airplane, one by one, in a way that more or less optimizes efficiency—something that seems banal to most of us but is way beyond the reach of most other species of apes.[2]

Cooperation is of course much more than simply not stepping on other people's toes. Humans everywhere reside in groups and engage in collective action that is beneficial to all, and it has been so for a very long time. From the earliest record of modern humans, we find evidence for col-lective hunting and for collective defense of the group, and it is very likely that early humans, like modern folk, engaged in many other such collabora-tive activities, from building shelters to pooling knowledge, and from assist-ing the old and the wounded to nurturing and protecting children. True, other organisms do provide help to conspecifics and can even create gigan-tic cooperative organizations like ant colonies. But this all happens between kin, as the ants or termites in a colony are all sisters. The unique feature of human cooperation is that it extends so easily to individuals beyond one's kin, and in large groups, even to individuals one does not really know at all.

Cooperation is a puzzle, at least at first sight, if you consider that indi-viduals are shaped to optimize their fitness.[3] In fact, philosophers and other thinkers had identified the problem long before we even had a notion of fitness. In most contexts, people generally benefit from cooperation. By joining a group and working as a coordinated team, we can hunt stags, or whales, or elephants, rather than just rabbits and mice. A share of big game can be much larger than a small prey. All participants benefit. But of course it would be even better, for each individual, to let others do most or all of the work, and still take a share of the proceeds. As Rousseau pointed out, one could abandon the collective hunt for a stag as soon as one chanced upon a hare, a much easier prey. And this is indeed a temptation that some cannot resist. But if that is the case, why doesn't everyone do just that? Which would make all cooperative endeavors unravel. The notion of genetic fitness rephrased this puzzle in more precise terms. Whatever genome makes you a cunning defector should be, it seems, a winner in competition with honestly cooperative genomes.

So why are humans (and some strikingly distant species like social insects) so good at something that eludes the rest of the animal kingdom? We usually tend to ascribe this to the civilizing and moderating influence of "society," but that is not much of an answer. Although it may seem to us that there is a society out there that is imposing norms on individuals, that is probably the least promising way of addressing the issue. What makes society do that? Society is an aggregate of people, and if some of them impose norms on others, that must be for some reason, which is precisely what we want to explain.

For a long time, the best information we could have about our disposition for cooperation and our sense of fairness was provided by moralists and novelists, and in a more systematic manner by pioneer social scientists like Adam Smith, who argued that cooperation was grounded in empathy, in the possibility for a human being to simulate the experience of being another individual.[4] Modern evolutionists would say that Smith was very much on the right track there. But until recently there was no clear way of showing how cooperation could have appeared in humans, and indeed how it could evolve in any species. We could describe human cooperative preferences, but we could not explain where they came from.

All this changed in the twentieth century, as economists and biologists found a way to formulate the question in a precise manner, as a result of introducing economic and game-theory models into evolutionary biology.[5] Competition and cooperation could be studied, not in terms of bluntly aggressive or peaceful instincts, but as different strategies, that is, moves that would result in higher or lower fitness in individuals, and thereby increase or decrease the frequency of whatever genes resulted in those behaviors. At the same time, economists were developing new techniques to investigate transactions in the laboratory, rather than merely trust the theorems of standard economic theory. Experimentally controlled economic interactions—markets, auctions, distributions of profits—would avoid the many confounding factors in the actual economy, and therefore illuminate the way humans beings reacted to specific kinds of interactions and incentives.[6] Finally, evolutionary anthropologists and psychologists joined this movement and investigated the social conditions and psychological makeup that would make cooperation possible, showing how cooperation occurred not by taming our spontaneous preferences but on the contrary as a direct result of our evolutionary heritage.[7]

Apparently Irrational Altruism

Are humans irrationally generous? This seemed to be the major conclusion from the first wave of research into cooperation. In particular, it seemed that people's behavior in economic games clashed with the optimization of expected utility—in other words, the prudent self-interest predicted by standard economic theory. In a Dictator Game, for instance, experimenters give a subject, the Proposer or Dictator, some money with the possibility of keeping all of it for herself or giving some of it to another participant, called Receiver, whom she never meets, about whom she may know nothing. The funds are then allocated the way the Proposer indicated. In the Ultimatum Game, the Receiver may accept the allocation indicated by the Proposer, in which case both receive what the Proposer indicated, or she may reject it, in which case no one receives anything.

In these experiments, most people turn out to be far more generous than economic theory would predict. For instance, in the Dictator Game, when participants could simply give nothing at all to the unknown partner and pocket the entire amount they received, they frequently give away half of the money or more. This is the case, too, in Ultimatum Games, in which Proposers, who should give as little as possible to maximize their gains, often offer half the money to the other player. The Receivers should in theory accept any offer, however low, as any amount is better than zero, but they frequently reject offers they find too miserly. Clearly, people are not following the principles of decision making expected from rational economic agents.[8]

These are not isolated findings, as hundreds of replications came up with similar results, including in different kinds of games. In "public goods" games, for instance, where people can choose to contribute to a common pool or hoard their gains, participants often avoid the tempting (and rewarding) selfish strategy.[9] And the results are not special to Western populations either. When Joseph Henrich and a team of anthropologists reproduced these game experiments in a dozen different societies, from foragers to agriculturalists and herders to modern industrial places, they found that nowhere did participants react the way economic theory predicts and recommends.[10] They almost invariably allocate more. People generally justify their unduly generous offers in behavioral games in terms of fairness, saying, for instance, that "it would not be fair" for them to take all the money, or that they refuse the miserly amounts offered to them in an Ultimatum Game "because it's not fair."

This so-called prosocial behavior, in early laboratory experiments, highlighted the puzzling nature of human cooperation, the fact that it seems to go against the logic of natural selection. Consider the frequently observed human behavior of pooling effort in collective action—and its experimental equivalent in public goods games. Like other traits, a disposition for cooperation varies between individuals, from very selfish to highly prosocial. If slightly more selfish organisms reap more benefits than others from interacting with others, they will have higher fitness, that is, will be more likely to pass on their genes to their offspring, including whatever genes drive their slightly more selfish than average strategies. So selfish behaviors should soon become widespread. So cooperation would stop. But cooperation does occur, so something else is happening.

Cooperation from Punishment?

Experimental evidence seemed to suggest that people were spontaneously motivated by a desire for generous rather than maximally beneficial allocations. Inspired by these results, the evolutionary anthropologists Rob Boyd

and Pete Richerson proposed a sophisticated interpretation of the origins
of cooperation. The main hypothesis was that people who cooperate, either
in laboratory games or in everyday life, are following social norms that include an aversion to inequality and a preference for prosocial behavior in
oneself and in others. Now the difficulty is to understand why this norm
would ever be transmitted. Boyd and Richerson argued that prosocial
norms could be stabilized, in human groups, by punishment. Indeed, there
is experimental evidence that people do try to punish noncooperators. For
instance, many participants in economic games choose to spend some of
the money they were allocated in order to diminish other players' income
when those players had not sufficiently contributed in previous rounds.[11]
This is often called "altruistic punishment" to emphasize the fact that people seem to spend their own money to enforce a norm that will benefit
others.[12] Mathematical evolutionary models showed that punishment could
ensure the transmission and stability of any kind of behavior. In this case,
punishment would have stabilized the existence of intense cooperation
inside human groups.[13]

As proposed by Boyd and Richerson, and further developed by other
anthropologists, the scenario assumed that some groups developed cooperative norms sustained by punishment. Because cooperation yields higher
resources, on average, than selfish defection, these groups could provide
better welfare to all their members. By comparison, groups with lower cooperative norms or less punishment would be less successful. Also, many
people would have migrated from weak reciprocity to strong reciprocity
groups, thereby contributing to the gradual disappearance of the former.
Note that this does not mean that low-reciprocity people perished, just
that the groups with weak norms fizzled out and lost their members—this
is cultural, not genetic group selection. Boyd and Richerson also showed,
using formal models, that such a situation could lead to a spread of the cooperative norms, as the higher-solidarity groups would absorb or conquer
the less cooperative ones. So humankind would have gradually changed
into a collection of ever more cooperative populations.[14]

But did it happen that way? Obviously, we have no direct evidence
for what occurred in the slow emergence of highly cooperative groups, a

process that probably began long before the appearance of modern humans.[15] But we do have evidence for modern humans' psychological dispositions, and for human behavior in very diverse cultures. Both the anthropological record and psychology experiments suggest that models of cooperation based on the results of early economic game studies may have been slightly misleading.

Consider the psychology first. In Dictator Games, an experimenter gives a participant some money, which now is hers to give away to some other person. Those are the rules. But do the participants' minds work on the basis of these rules? True, the participants have read all the instructions and can repeat them to experimenters. So they "know" that they can keep all the money, give it all away, or do anything in between. But most psychologists would argue that this does not show that the participants' behavior is actually driven by those rules. Indeed, variations on the Dictator game show that something else is going on in the participants' minds. In some experiments, instead of being given free money, people "earn" it by performing some task, such as building a toy with Lego bricks or solving anagrams. Now people who have earned their money that way tend to be much less generous than in the standard version of the game. They want to keep their earnings. In Ultimatum games, too, those who worked for the money also offer much less.[16] This of course makes intuitive sense. Players who worked to get the money clearly feel more entitled to it than if it was merely handed over to them by a complete stranger for no clear reason. It seems natural that they would want to keep the money and that we should accept any low amount they care to give us.[17] So, going back to the standard version of the Dictator Game, even if you were told that the money was yours, some intuitive system that regulates ownership probably suggested that it really was not yours at all. Even when an experimenter tells you that your ordinary intuitions need not apply, your mental systems apply them anyway—just like the warning that "all the events and persons described here are fictional" has never stopped our emotional systems from engaging with the characters in films and novels.

Another complication was that apparently altruistic punishment—spending money to punish those who did not cooperate—seems to be less than really altruistic. When you punish the miscreants, even those who did not harm you, it may seem as if you are working to uphold the common norm of fairness. But, in some intuitive way, you may behave in a manner that shows you as a good, norm-abiding cooperator—you burnish your reputation—and you may also be sending a signal to deter anyone tempted to exploit you. Experiments show that both motivations are at play here. Experimenters can manipulate the level of anonymity in public goods games, which, if the punishment model is correct, should not make any difference, as people are mostly motivated to punish norm violations. But the manipulation does have a clear effect, as people punish much less when no one can notice.[18] Also, when people play economic games, they spontaneously assume that the behavior of an individual toward other players predicts his behavior toward them. When the situation designed by experimenters removes this prediction, people stop punishing third parties—which suggests that they did it as a deterrent.[19] So punishment of third parties, when it occurs, is motivated by many factors, but the desire to maintain cooperative norms is probably not among them. Indeed, players who do not know whether or not their punishing attitude brings benefits to the group punish defectors anyway.[20]

In anthropological terms, too, the punishment theory seems difficult to maintain. The model implied that one would punish an individual for not cooperating with a third party even if one did not really stand to lose—that is called third-party punishment. But it turns out that there is very little third-party punishment in real social interaction, at least in small-scale societies. People in most groups will punish norm violators, but they do so mostly when they have suffered from the violation. In fact, anthropologists report that in most small-scale societies there is very little active punishment, let alone costly punishment of nonreciprocators. Rather, people just deplore the defectors' bad character and prefer to interact with other, more cooperative types. As an illustration, a study of collective action in a small Tsimane group in lowland Bolivia showed that people often shirk their communal duties, for example, helping repair wells and bridges, without much consequence.[21] This is in fact typical of small societies, where people do retaliate against offenses, and mobilize their kin and friends to do so when appropriate, but rarely bother to punish third-party violations, as this is of no great benefit, and potentially costly. In general, to the extent that people are not directly affected, they prefer to ignore violators.[22]

Obviously, there is a great deal of third-party punishment in modern societies. You do not pursue thieves and muggers yourself but trust the police and the justice system to mete out appropriate sanctions. These institutions do provide third-party punishment. That, however, is of course not a process that is relevant to human evolution, as these institutions only appeared very recently in an evolutionary timeline. And they only appeared when the size of the communities lowered the cost of third-party enforcement. Paying your taxes to support law enforcement is much cheaper than hiring your own police force.

So, if altruism enforced by punishment is not the explanation, why do we have human cooperation? And why does it manifest in such generous behaviors as observed in some economic ages? Maybe considering very distant species could provide some insight about the way mutually advantageous interaction might evolve.

The Wisdom of Fish

In some reef environments, one can observe an apparently stable and cooperative interaction between so-called clients, large fish that need to get rid of parasites on their skin, and their cleaners, much smaller members of species like the cleaner wrasse. Cleaners nibble parasites off the scales of the much larger clients, an exchange of food for hygiene. In some species, there are established cleaning stations, local spas where clients make occasional stops and cleaners stand ready to service them. This may seem like an unproblematic exchange of services, as the large clients need parasites removed, while cleaners need food. But it also comes with conflicts of interests. In particular, cleaners feed on the parasites, but they much prefer the clients' mucus, which the latter need to keep intact, as a protective layer. So the interaction offers opportunities for defection from cooperation.[23]

Formulated as a Prisoner's Dilemma, this kind of interaction should quickly unravel, as defection may be advantageous in the short term. In particular, the small cleaners would bite off as much mucus as they could from a client and scamper off. That is not the case, however. Cooperative exchange prevails, and it was stabilized by natural selection. How is that possible? As described by biologist Redouan Bshary and his colleagues, several factors explain this. First, both cleaners and clients are in competition. For each client there are many potential cleaners, which themselves enjoy a choice of clients. So if a partner does not cooperate, one can always find a replacement. Second, clients can punish misbehaving cleaners, by shaking them off and then avoiding them. Experimental studies show that cleaners do learn their lesson and mend their ways as a result of moderate punishment.[24]

So these apparently simple organisms have found a way to engage in mutually beneficial transactions. Obviously, no one is suggesting that the fish compute all these contingencies explicitly. They do not need to, as evolution by natural selection has provided them with the right preferences and the right reactions to situations of cooperation and cheating. Genotypes that resulted in lesser discrimination, as well as unduly cooperative or dishonest strategies, could not spread in the gene pool as much as those favoring mutually favorable interactions. The central feature that accounts for this successful interaction clearly is the possibility to choose partners, ditch defectors, and stick with cooperators.[25]

Given the advantage of this form of interaction, this would seem to be a good example to follow for human beings . . . except that they already behave like that, which may explain how we evolved to engage in intense cooperation. This comes of course with the crucial difference that cooperation between humans is in our case a form of intraspecific mutualism, where individuals cooperate with fellow members of the species. But it may be the case—indeed, the evidence strongly suggests it—that we built cooperation between unrelated individuals because we were faced with the same circumstances that allow cleaner-client interaction, namely, a choice between possible partners, a knowledge of each individual's past interactions with others, and some graded punishment for occasional defection. The possibility of partner choice would have created a market for cooperators, as individuals differed in their offers and adjusted these offers to a potential partner's preferences, leading to advantageous outcomes.[26]

This is obviously very different from classical models of cooperation, based in particular on the Prisoner's Dilemma or on reiterated Dictator's Games, in which one person is faced with only one potential partner at each stage, and the only question is how to avoid defection from that partner. In such situations, there is a great benefit for each party if both cooperate, but an even greater one for the cheater faced with a cooperator. The equilibrium in such games is of course to defect, and as both partners can reason by backward induction, they expect the others to defect and will themselves defect.

But human social interaction never consisted in such dilemmas.[27] Humans evolved in groups where they could offer and receive cooperation from different individuals. Humans also evolved to have an intense interest in the affairs of others, such that information about your behavior with various individuals is usually broadcast far beyond the interested parties. So one may benefit a great deal from having a reputation for honest, mutually advantageous behavior.[28]

Cooperation for Mutual Benefits

The existence of partner choice explains otherwise puzzling features of human cooperation. For instance, anonymity is a very difficult thing to process for human minds. In the cross-cultural replications organized by Joseph Henrich and his colleagues, getting participants to think of the games as really anonymous proved exceedingly difficult.[29] Participants in classical Dictator Games are told that Receivers will not know their identities. That is crucial if we want participants to behave in ways they think optimal, rather than worry about retaliation or reputation. People recruited for these studies may well state that they understand the procedure is entirely anonymous, but it is not clear that all the relevant mental systems are actually working on that assumption.[30]

Also, in the social environments in which we evolved, individuals would interact with each other again and again. Indeed, even when given explicit instructions to the contrary, participants in economic games spontaneously assume that interactions will be repeated—which of course affects their behavior, as they are motivated to cooperate with individuals who may have the opportunity to return the favor.[31] This may be why people are usually generous in one-shot encounters. They give some of their money in Dictator or Ultimatum Games. In less contrived situations, people often tip in restaurants that they will not visit again. Generous behaviors of this kind may seem difficult to explain in terms of narrow self-interest, but in the evolutionary context of small-scale groups they would constitute a first step in the construction of mutually beneficial arrangements. In such environments you can lose out, not just by choosing the wrong (noncooperative) partners—that is the cost that traditional models focused on—but also by failing to interact with the right ones, missing out on an opportunity for long-term cooperation.[32]

If there is some choice of potential partners, there is certainly little advantage in engaging in third-party punishment of people who violate the local norms of cooperation. In those cases where you do encounter defectors, the simplest option is often to withdraw from any further interaction with them, and to seek other people with better dispositions.[33] This of course is a form of punishment—the miscreant is not really much affected in the short term, but it is also rather cheap for the punisher. Note that this is a strictly individual strategy. That is, you are not withdrawing cooperation from that person in order to benefit the group, or to make certain norms better, or as people sometimes say, "as a matter of principle," but principally for your own interest.[34]

The partner-choice framework also makes sense of more subtle aspects of punishing defectors. Fish that engage in mutualism have gradations in punishment. Clients faced with voracious cleaners that bite a bit more than they should can shake them off, decline their services, or even chase them. Graded punishment helps the client avoid exploitation, and it teaches the would-be exploiter a lesson. In formal terms, this enhances the possibility of mutually beneficial interaction, because once a past cooperator has tried to get a bit more than it should, and has been punished for it, that individual knows how not to go too far, so to speak. One would predict that, all else being equal, it is better to interact with such an individual than with an unknown new partner. That is indeed what happens in economic games, when participants can interact with multiple partners, reward cooperation, punish defection, and acquire information about other people's past interactions. People prefer a previously punished partner to a new one about whom they have no information.[35]

But what about fairness? Humans do not just cooperate, in the sense of participating in collective action and reaping the benefits of these mutually beneficial arrangements. They also try to maintain fair allocations, and react most vigorously when people try to take more than a "fair" share of the proceeds. Consider the simple problem of allocating the benefits from hunting, or digging a well, or putting together a lemonade stand. It seems intuitively obvious that the benefits from such endeavors should be divided equally between participants who contributed equally. The proceeds from the stand should be split between the investors, and the water from the well should be accessible to all those who helped. In case there are unequal contributions, it seems just as obvious that each participant's share should be proportional to her effort. Those who harpooned the whale, a more dangerous and crucial contribution than steering the boat, should have first choice in the distribution of the meat. Indeed, that is the way collective hunting and all manner of collective action are organized in most human societies. Experiments show that even three-year-olds have the intuition that rewards should be proportional to contributions, in places as different as Japanese cities and the camps of Turkana nomads in Kenya.[36] Obviously, it does happen that people take more than their share—but that is universally considered exploitative, and people are eager to avoid or shun individuals who do that.

This intuition about proportional allocation may stem from partner choice. Consider, for instance, that you propose to engage in some joint operation with a partner, where both will provide the same amount of effort, say, make lemon juice for a lemonade stand. How should you divide the benefits from that operation? Most humans have a strong intuition that the benefits should be evenly split. But why not propose an unfair allocation that is better for you? You may offer, for example, an exploitative 80/20 split. If your collaborator has a choice of potential partners, all competing for some such cooperation, many of them will offer more favorable terms than just 20 percent of the proceeds. So it is unlikely that you will find anyone willing to accept such unfair deals, if they have other options. Should your partner become greedy and insist on an exchange that would exploit you . . . you can turn to others, until you reach the best you can get, that is, about half of the profit. This feature of partner-choice models, which corresponds to human intuitions about "fair" distributions, can be formally demonstrated from the mathematical properties of partner markets.[37] It also corresponds to what happens in the field. In societies of hunter-gatherers there is a clear correlation between giving and receiving—what goes around comes around *in proportion,* which is the crucial point.[38] So a major difference between forced-choice and market models is that the latter, contrary to the former, explain not just the fact of cooperation but also the amount of cooperation.

All this converges to suggest a plausible minimal scenario, to explain how cooperative dispositions were stabilized in human evolution. Humans are special in that they extract a great part of their welfare from others, through both kin solidarity and collective action (hunting, shared parenting, group defense, and the like). Collective action can emerge and become extremely advantageous if individuals, first, have the cognitive capacities to keep track of different individuals and to keep a record of past interaction with them. Second, the scope of exchange becomes vastly broader once individuals can get information about interaction between others. Unlike fish, humans can and do broadcast information through communication, so that people's past behaviors are almost certainly known to all who might interact with them. What biologists call reputation is really the circulation of social information that we know is present in all human groups, and is especially intense in small-scale groups.[39] Third, individuals must be able to inflict graded punishment on noncooperators, from simple avoidance, to spreading information about their behavior, to seeking direct retaliation. All these capacities are characteristic of human beings, so that emerging cooperation did not require a sudden leap in capacity, caused, for instance, by a large mutation. Also, these capacities are such that they could be gradually honed by selection—they do not consist in all-or-nothing strategies. So a population where there is limited cooperation can favor the spread of genotypes that favor more refined cooperation skills.[40]

Sharing and Trading

Armed with these capacities for mutually beneficial cooperation, our ancestors engaged in all manner of collective action, like collective hunting, group defense, or shared parenting. But they also engaged in two forms of economic activity that are uniquely human. They practiced communal sharing, and they engaged in some form of trade with strangers.

The most salient aspect of cooperation in small-scale forager economies is communal sharing, in which individuals pool resources with the entire group. Sharing is found to some variable extent in all human groups, particularly in food provision, and seems crucial to social interaction in foraging groups similar to those in which humans evolved. The issue of why and how sharing developed in humans became an important issue for evolutionary anthropologists and psychologists. Food sharing removes resources from the individual, which is mysterious if you believe that individuals always maximize their own welfare. And sharing goes far beyond one's immediate kin and dependents. Did this mean that in our ancestral conditions we evolved to be unconditional altruists? In early anthropological accounts, foragers were indeed described as moved by a generous ethic of indiscriminate sharing. The notion was that there was virtually no private property in forager groups, as everyone shared the products of their activities with everyone else.

But communal sharing is more complicated than that. First, in hunter-gatherer economies, resources are not all shared in the same way. People share gathered and extracted plant foods (berries, tubers, roots, leaves, and so on), but only with close kin. Game, by contrast, especially big game, is generally shared at the level of a group. Everyone will probably get something, but the hunters themselves keep a greater share than others, and among hunters the ones responsible for the final assault or kill often get an even greater share.[41] The difference in sharing between plants and game makes sense as a form of insurance. Most gathered resources, like roots, tubers, nuts, and berries, are of low variance, so that any individual can expect to extract neither less nor more than others, and tomorrow in much the same way as today. By contrast, the returns from hunting vary greatly, with successful expeditions few and far between. So it makes sense for people to share the game they bagged, as an assurance that they will in return receive some share of resources when they are out of luck. Sharing also makes sense given the marginal value of food units, which decreases steeply for large prey. Big game in particular comes in units that are much too large for an individual and his close kin. If you do not share your big prey, it will simply rot away, and you will look selfish. So sharing creates obligations and prestige at a low cost.[42] Most important, sharing is most often conditioned on past or expected reciprocation. Even where there is a norm of unconditional sharing, those who give more freely also end up receiving more.[43]

So communal sharing is not based on a simple urge to favor others. The intuitive system that guides our distribution preferences takes as input information describing (a) the resources provided, (b) the identity and behavior of various individuals involved in producing the resources, and (c) the identity of people who are claiming a share of the resources. The same system produces intuitions about a desirable distribution, adjusting the share attributed to each according to her contribution, the nature of the resource (for example, game versus plants), and her connection to the contributors.

Beyond sharing, there is of course trade. In all known human groups, even with the simplest foraging economy, people also engage in trade, exchanging services, objects, or favors. The objects may consist in tools, clothes, ornaments, toys, medicines, and much more. All these, as well as various services, could be traded within a group. But there was also some long-distance prehistorical trade between groups, particularly in precious commodities, like obsidian in Europe, Africa, and the Andes, as well as materials and substances that could not be sourced locally, salt in particular. Prehistorical Europeans conveyed goods along the Danube and other large rivers. Cowrie shells from the Indian Ocean found their way to China and West Africa, where they served as currency, while obsidian from Mexico was used in Mississippian cultures.[44] The archaeologist Colin Renfrew listed the many paths that may lead from straightforward home-based production to modern mass markets—including local exchange with kin and group members, communal redistribution in a group, central marketplaces, centralization by chiefs, middleman trade, emissary trade, colonial outposts, and so forth.[45] Different kinds of goods may have circulated along these different kinds of "commodity chains," and in many documented cases a specific good would circulate along several of them.[46]

So trade occurred, but until recently it affected only a small part of production. Humans gathered plants and hunted game, and then started cultivating small gardens and domesticating plants and animals, but for most of prehistory they consumed most of what they produced. Why did trade, an activity that humans beings can and will readily engage in, occupy a relatively small place in the economy? Part of the answer is that there were not that many goods to trade. Foragers do not, for instance, usually produce a large surplus of food, and that surplus could not be preserved in most cases. Even if that was possible, the low population densities in human prehistory would mean long trading distances, another obstacle to efficient trade.

The Psychology of Exchange

However limited compared to modern commerce, prehistoric trade demonstrates that humans had at some point developed the specific psychological capacities that make it possible to exchange goods and services for other goods and services, a form of interaction that is very rare in nature.

The capacity to trade seems to be uniquely human. To us, trade seems both transparent and rational. What could be simpler than giving away what you do not really need, for what you want but do not possess? The fact that something is transparent or self-evident does not mean that it is simple. It indicates that the systems that make it possible are well designed, and designed to do their work away from conscious inspection. So if we try to look under the hood of trade, so to speak, what mechanisms can we see?

First, obviously, agents who trade goods or services must be able to measure their respective utilities and infer that gaining one is worth losing the other. This is both entirely self-evident to human minds, from an early age, and almost unfathomable to organisms of most other species. We immediately see that barter can solve what economists called a coincidence of needs—as when I really want bread but have too many sausages, while you are in the exact opposite predicament. Just describing the situation, for human minds, immediately suggests the solution. And we intuitively appreciate that the transaction makes both parties better off. But this is a special cognitive adaptation, which is why it is so rare in nature. True, there are many cases of reciprocity in the animal kingdom. Grooming among primates is a familiar example. In the same way, vampire bats store blood and regurgitate it into the mouths of companions that did not succeed in finding prey, with the expectation that the companions will return the favor at some point. But these are all deferred exchanges of the same good or service. They rarely if ever include different kinds of objects, trading, for example, grooming for food. There are also instances of quasi-exchange in the context of gifts for sex, but that is generally restricted to only one type of service or object.

By contrast, generalized trade requires, precisely, the ability to consider a large variety of goods as possible items of exchange and, crucially, the capacity to evaluate the value of one good against some amount of another good by a common measure of utility. Neuroeconomic studies illuminate the neural systems involved in valuation and decision making, in how individuals compare the expected utility of various choices, for example, prefer one kind of good to another, or some amount now as opposed to more later. They show that similar circuitry and processes can be observed in close primate species, like monkeys and humans.[47] Indeed, some nonhuman animals can be trained to trade and even use currency tokens. But such behaviors clearly do not belong in these species' evolved repertoire, while they are ubiquitous in humans.[48]

A second set of skills is the capacity to represent ownership, in a flexible and subtle manner that allows exchange. There is of course no trade without clear knowledge of whose goods are being passed along. Our representations of ownership are often highly intuitive (it feels as if we simply know what it is to own something) and remarkably difficult to articulate. The world over, human beings can confidently assert that certain individuals own certain things. All natural languages can express the special connection between agents and things. Also, in human cultures one finds a distinction between ownership and mere possession. The fact that I am driving the car does not make me its owner, especially if you know that I forcibly took it from someone else.[49] There is no known human group where people fail to make that distinction. Finally, strong emotions and motivations are everywhere associated with representations of ownership. Stealing other people's goods triggers anger and a punitive sentiment. Possession of prized goods is the source of pride, satisfaction, or envy.

In humans, ownership intuitions develop very early, and they include an expectation that the first possessor is the owner—as the saying goes, possession is nine-tenths of the law. Even in very young children, however, the distinction between possession and legitimate ownership is crucial. Some people may be holding an object and using it without being considered owners, and owners may not be in possession of what they own. This presents no conceptual difficulty for young children. What makes ownership special, for them and for adults, is the history of a person's connection to a thing. For instance, young children share the intuition that the person who extracts some resource from the environment is the owner. They also assume that transforming an object, such as turning a lump of clay into a sculpture, makes one the owner of the finished product, more so than the original possessor of the clay.[50] Naturally, people also have explicit beliefs about ownership. But experimental research shows that these beliefs are often vague and confused, and occasionally incoherent. People, for instance, state that one cannot literally own persons, before being reminded of slavery, or that ideas are not property, before songs or movies are mentioned.[51] The fact that our explicit conceptions are often incoherent suggests that they are not the source of our ownership judgments, which are in fact governed by the intuitive ownership system. Here, as in other domains, reasoning comes after intuition, as an attempt to explicate or justify it.[52]

Finally, our capacities for social exchange include systems dedicated to detecting free riding and cheating, which occur when individuals manage to extract a benefit from a transaction without paying the associated cost. Early in the development of evolutionary models of cognition, the psychologist Leda Cosmides reasoned that humans have probably evolved a specific inference system that would identify information of the format "benefit received, cost not paid" and trigger the appropriate threat detection. Indeed, experiments showed that detecting free riding was automatic and very specific.[53] People usually find it difficult to identify what information would confirm that a rule is actually being followed—for example, "if the folders are green, they contain approved applications," would an approved application in a red folder disprove the rule? Most people's responses are wrong, assuming against logic that red folders only contain nonapproved applications. But if the rule is formulated in terms of benefit and cost, for example, "if the drink is alcohol, then the customer is over eighteen," then it creates no such difficulty. Psychologists found similar results among American college students and Shiwiar hunter-gatherers in the Amazon.[54] Our free-rider psychology is not just about the combination of benefits received and costs not paid. If that were the case, people would be motivated to punish those who cannot contribute, for instance, because they are too young or too old, or temporarily unable. Conversely, we would try to punish those who took benefits by accident, without realizing it. But experiments show that entirely unconscious systems sort out these different cases and focus the emotional reaction, as well as avoidance response, on those individuals who deliberately take more than they should and avoid contributing.[55]

A Template: Embedded Social Exchange

These three cognitive systems—utility equivalence, ownership, and free-rider detection—contribute to what could be called a template for exchange, an abstract description of what our mental systems expect a transaction to be like. Actual transactions may not correspond to that template. In fact, in the modern world many transactions differ from these expectations. Nevertheless, the experimental evidence suggests that this template is effectively part of our spontaneous expectations about trade, and probably has been part of our mental equipment for a long time.

For one thing, humans spontaneously expect exchange to take place between identified agents, as this is necessary for partner choice. We need to keep the identities of cooperators and defectors distinct, and we also need to maintain records of past transactions with each particular individual, as we could not otherwise evaluate what to expect from her and what she expects from us. Also, identification is of course required in order to receive or broadcast information about who is cooperating with whom, and how much—and human partner choice depends on reputation. Also, humans spontaneously expect and prefer transactions to be voluntary. That is manifest in their emotional aversion to exploitation, to situations in which stronger individuals or groups can impose intuitively unfair conditions on weaker partners.[56]

In our exchange template, we also assume that individuals who met will meet again, and that mutually profitable interaction occurs in the context of repeated transactions. In a situation that allows partner choice, but where trade is mostly limited to short-distance exchanges and groups are small, it is advantageous to carry on trading with the same individuals. This allows partners to exchange favors, which constitute a form of insurance against exploitation. The person who usually trades arrowheads for your honey will not try to jack up his price if he knows you have run out of arrowheads, because he wants to keep you as a partner in the future, and this of course works in both directions. This pattern of reiterated exchange, with occasional favors thrown in, implies that transactions have no fixed horizon, no clear limit after which nothing more will be expected. This is probably why pure one-shot barter, where partners exchange goods on the spot and expect no further interaction, only played a limited role in the economies of small-scale societies. It occurred mostly at the periphery of exchange, typically with strangers rather than members of the band or tribe.[57]

As a result of evolved preferences, our spontaneous template for social exchange combines information that is economic in the strict sense of the word, concerning goods, their amount, their value, and their prices, with all sorts of information to do with the identity and reputation of persons. As many anthropologists have pointed out, economic activity in small-scale societies, and by extension during most of human evolution, did not and does not take place in isolation from other aspects of social interaction. Transactions affected not only the welfare of the agents, what they gained or lost, but also their reputation, their social standing, the nature of their relationship to exchange partners, the extent to which they could rely on others, the cohesiveness of the groups they belonged to, and so on. The clear separation of economic exchange from other aspects of social interaction is a recent by-product of market conditions.[58]

So humans evolved a capacity not just for trade but for repeated transactions with known partners, with carefully monitored mutual goodwill.[59] In the past, anthropologists were divided concerning the best way to understand economic processes in societies without mass markets. While some insisted that the standard tools of economic theory should be used, others emphasized the additional factors described here, such as the importance of reciprocation, social investment, and reiterated trust, what could be called embedded trade, that is, transactions where such social factors are as important as direct utility in people's decision making.[60] These anthropologists were quite right to emphasize that trade in a small-scale society is rarely only about trade, so to speak. But they erred in assuming that these additional factors, these considerations of goodwill and reputation that people spontaneously include in their appraisal of a transaction, were a matter of arbitrary norms, of "culture," so to speak. That is of course insufficient. People can learn local norms because underlying principles about partner choice, reputation, and fairness are part of our evolved learning systems. They constitute an exchange psychology that is unique to human beings and accessible to any normal human mind, in which trade is made possible and modulated by a whole variety of moral emotions and motivations, including gratitude, goodwill, envy, spite, and outrage at exploitation.

The Psychology of Commons

Our social exchange psychology also allows us to engage in complex forms of cooperation that go well beyond communal sharing or direct trade between individual partners. For instance, many resources in the world constitute commons, as the goods are rival (the more one uses them, the less is left for others) but also nonexcludable (it is difficult to stop people from using them). An example of a commons is the water from a river. All the farmers can take some water to irrigate their fields, but if one farmer takes too much, all the others will suffer. It is difficult to turn the water into a private or club good, that is, to exclude some from access to it. Anyone who is near can throw a bucket into the river and help herself. Another example would be hunting grounds. If some hunters overdo it, there will be much less available to everyone else. Again, it is difficult to exclude people from hunting—anyone can show up with a gun and help himself—unless one hires guards, at a high cost. The economic term is derived from the practice of having a "common" in English villages as a communal field where everyone's livestock could graze. This created the obvious possibility of overgrazing, of people trying to take advantage of this free resource and therefore deprive everyone in the long run. Indeed, the notion became familiar outside economics because of Garett Hardin's famous description of the "tragedy of the commons," arguing that such overgrazing (and in general overuse because of selfish motives) was inevitable in common-pool resources.[61] As time went by, all commons would be overexploited for short-term individual profit, leading to long-term collective loss. The argument was very successful in economic theory, leading to all sorts of formal models on the likelihood and speed of overgrazing.[62] It also became a familiar item in discussions of policies, convincing many people that either privatization or state control would avoid the tragedy of overuse and depletion.

Starting in the 1980s, however, many social scientists began pointing out that the tragedy did not seem to occur, or a least not with the inevitability predicted by the model. Indeed, in many situations people managed common-pool resources for decades or centuries with little or no overexploitation. How did they do it? As the answer could not be provided by formal models but required detailed documentation from history or ethnography, many economists left it aside for a long time. Among the exceptions were Elinor Ostrom and her colleagues, who studied and documented such cases as irrigation systems or fisheries, trying to extract general lessons from the many successes and occasional failures of commons management.[63] In this domain, economic history provided a natural experiment, as people over the centuries and in many different places have tried all sorts of institutional arrangements, only some of which seem to lead to successful and stable systems. These obviously require rules that bind the users, but what kinds of rules actually work in the long run?

One condition is that the rules should clearly define the set of appropriators, that is, who is a legitimate user of the resource. This is not very easy in some cases. The set of fishermen on a small isolated coast may be the most natural set of users for the local fishing stock, but how does that extend to the situation of long-haul ships operating in the open seas? The natural set of users of a water basin consists of the farmers whose livelihood depends on irrigation, but that may change with a transition to an industrial economy that also requires water but is generally not locally grown. Also, commons work when the local appropriators themselves design the rules. But the most important conditions for success consist in efficient monitoring of people's actual use of the resource. In particular, the use of commons should be monitored, if not by the users themselves then at least by officials directly accountable to them.[64]

Commons are important in demonstrating how our psychology of exchange allows production to scale up, very far beyond the limits of small-scale societies in which it emerged. Ownership psychology provides intuitive criteria for limiting the use of a common to a particular set of agents. Communication and memory allow us to acquire information about other users' behavior. Cheater detection provides the motivations for deterring and occasionally punishing those who overuse the resources. The expectation that the same agents will be interacting in many future rounds offsets the temptation for short-term, selfish overuse of the resources. Graded punishment allows users to maintain a level of honest cooperation without the need for costly enforcement. So the psychology of small-scale exchange can help us manage the use of water from a river or fishing in an estuary. But what happens when trade scales up to include a nation or the entire world?

Complexity from Cooperation

The largest domain of human cooperation is generally invisible to all of us, or not understood as being a form of cooperation at all. I refer of course to market transactions. Every single person probably performs thousands of exchanges every year. That itself is not invisible—but what is difficult to see is how trade is a form of cooperation. We all recognize that trade is the most peaceful and efficient way of acquiring other people's goods—after all, what you cannot trade you have to beg, borrow, or steal. Market exchanges are cooperative because transactions are mutually advantageous, as any elementary economics textbook will state in its first pages, pointing out that transactions occur when bakers prefer the money to the bread (otherwise they would just keep their bread), while customers prefer the bread to the money (otherwise they would not part with their money). Exchange benefits both parties, as both get more than they had before.

The cooperative interaction that is trade naturally tends to expand, including more and more individuals in complex webs of cooperation and making them more prosperous as a result. One of the engines of this expansion is the division of labor. Adam Smith is usually credited for clearly formulating its importance for the creation of value. His famous example of the number of people it takes to manufacture a simple pin remains a classic demonstration of the advantages of division of labor. By specializing in one kind of operation, each agent can do more and do so better than if she had to switch tasks. But this effect had been recognized long before Smith explained it systematically. Already in fourth-century-BCE Greece Xenophon observed that "in small towns the same workmen make couches, doors, ploughs, and tables . . . so they cannot make [any of] them well," while in large cities you found specialists of men's or women's shoes, and indeed shoemaking itself was divided among cutters and stitchers.[65]

Xenophon mentioned the higher quality of goods, and Smith added that they came at a lower cost. But there is another dynamic quality to the division of labor, which is that it ushers in comparative advantage, as David Ricardo pointed out a few decades after Smith. For example, even if Sid, a successful writer, is faster at fixing software than Doris, his computer technician neighbor, it may still be advantageous for Sid to get Doris to do his computer maintenance. If he spent that time doing it himself, he would miss hours of writing that actually yield more revenue than the amount he pays Doris. Or, to use the right term, his opportunity cost (the price you pay, when you do something, for not doing the next best thing you could have done) would be higher than what he pays Doris. So division of labor favors further specialization, in a virtuous circle that expands trade and creates more value.[66]

Obviously, all this can only happen in situations of relative peace and security. For much of human history, most people were kept away from trade, apart from local exchanges, because of many limitations that made transaction costs, that is, all the costs involved in trying to exchange, prohibitively high. There was no impartial third party to enforce contracts. The only guarantee that people would fulfill their part of a contract was either to exchange there and then, in the form of barter, or to exchange with kin, or to be formidable enough that one's partner would not defect. Also, information costs were high. Beyond a small-scale circle of personally known partners, one could not really have much information about potential partners and products. Given these conditions, most trade took place along networks of sharing and reciprocity based on kinship, because relatives are people about whom we have reliable information and on whom we can exert some pressure.[67]

Later on, with the development of agrarian societies, trade was still made difficult by exploitative despots and warlords. When a king has the right to confiscate private property as he wishes, or to exact any levies he fancies on the production and circulation of goods, few transactions will occur. And when warlords threaten the population's security, very few people engage in trade. That is what happened to Western Europe after the fall of the imperial institutions that guaranteed the Pax Romana—the possibility for people to travel (mostly) unharmed across the empire. Very soon, after each Barbarian conquest of a Roman province, trade started to wither away as travel became risky. This marked the end of the cosmopolitan times of the late empire, when Italians consumed large quantities of African wheat, Spanish wine, and Greek oil, not to mention tools and artifacts from the far corners of the world.[68] In the same way, trade and industry in modern-day Africa are still hampered by widespread insecurity, which makes it very expensive to store and transport goods.[69] As well as the security of goods and persons, prosperous trade requires relatively secure property rights, and some way to secure and enforce contracts, which requires (again, relatively) efficient and uncorrupted judicial institutions. This does not necessarily require an expansive state. Medieval and early Renaissance traders for instance had their own associations to allow for trade between trusted partners, and shut out the others.[70]

The expansion of trade created a new challenge for human minds. Our dispositions for exchange developed in the context of small-scale groups, where there is a lot of information about different individuals' contribution to production. People in a foraging group know who hunted the antelope, who helped them—and these individuals are among the ones who claim a share of the game. But in mass-market societies we generally have very little information about who contributed to what, because production processes are vastly too complex for that. To reprise a famous example, even making a simple pencil requires the cooperation of hundreds of specialists, many of whom have no idea that they are participating in the construction of a pencil, and none of whom is motivated by a desire to provide the end consumer with that instrument.[71] What is difficult to fathom, for human minds and their social exchange psychology, is not the numbers of agents involved but the fact of coordination, that is, the apparent miracle that makes different, unrelated, and not particularly altruistic agents provide (more or less) what is needed, at the time it is needed and in sufficient amounts, for people to make and buy pencils, pianos, computers, or sausages. This massive phenomenon of coordination is unlike any other social interaction known to human minds—which is why our cognitive systems have great difficulty making sense of it.

Do We Understand the Economy?

Trade expanded to the extent that it now involves billions of individuals in impersonal exchanges, and most goods consumed are produced outside the home or the local community. Our welfare, and the relative positions of people in society, depend on the aggregation of an immense number of single transactions, most of them impersonal exchanges with no expectation of repetition. This is not a situation for which we have appropriate intuitive capacities, as these conditions appeared recently in the history of humankind. By what process can human beings judge the fairness, or desirability, or even the efficiency, of what happens at the scale of mass markets? How did humans develop a new kind of knowledge, to understand what happens when thousands or millions of individuals act in coordination, driven by their individual pursuit of profit?

The short answer is that they did not. Even though humans have been engaged in long-distance trade for millennia, and even though global commerce emerged centuries ago, this did not radically alter our psychology of exchange, which is why people's behavior in laboratory economic games is often so close to what would be optimal in small groups of familiar partners. That is also why evolutionary psychologists can make sense of many otherwise puzzling aspects of consumption in modern societies.[72] Markets may be all around us, they may coordinate in an efficient manner the behavior of millions of unrelated agents, but we still think about the economy, most of the time, in terms of our evolved social exchange template.

Most people hold beliefs about, for instance, rents, wages, unemployment, and welfare or immigration policies, as well as mental models of interactions between different economic processes, for example, inflation, unemployment, whether foreign prosperity is good or bad for one's own nation, whether welfare programs are necessary or redundant, whether minimal wages help or hurt the poor, whether price controls make prices go down or up, and so forth. I am referring here to the views of laypeople, distinct from what experts may say about these various economic realities, what could be called "folk-economic beliefs."

These views about the economy are crucial to politics. In modern democracies, people identify with political parties, and adhere to their policy proposals, mostly on the basis of a general view of the economy. They evaluate specific policy proposals on the basis of possible economic effects. Will rent controls keep rents low? Will taxes reduce inequality? Will pollution increase unless we regulate emissions? So it is important to document these folk-economic beliefs and try to understand the implicit theories of production and exchange they are based on.

Surprisingly, there are not that many systematic studies of folk economics. But what we have is enough to suggest that people's views on economic processes are not random, that they are not simply the outcome of media influence or political propaganda, and most important, that they often clash with professional economists' understanding of what happens in mass-market societies. Indeed, to the extent that social scientists studied folk-economic beliefs, such studies were often prompted by the fact that so many people seem to adhere to views that economists find either false or misguided.[73] So, in what way do people seem to get the economy wrong?

One often-cited belief is that the creation of value is a zero-sum game. So if some individuals or groups get larger incomes or wealth, this implies that others become worse off. In other words, there is a fixed pie, as the phrase goes, that can be divided in many possible ways but cannot be enlarged. For some to have more, others will have to lose. This is so common in so many political debates that the point hardly deserves more detailed illustration. Against this, economists would point out that while zero-sum games are an apt description of some interactions, like warfare or candy distribution, by contrast the economy as a whole is a positive-sum game. Otherwise, there would be no general increase in prosperity, especially not on the dramatic scale observed in the last centuries. If everyone became more prosperous, the pie must have become larger. Another, related belief is that the wealth of nations is also a zero-sum game. Many Americans think that if China becomes richer, that will be bad for the United States. Against this economists would of course argue that richer Chinese people will have more opportunity to buy more American goods. It is in everyone's interest that others get richer, especially your customers.

Another common assumption is that prices are determined only by bargaining power. This is a highly widespread assumption, which most economists find terribly misguided and misleading but nonetheless crops up in many discussions of prices. The notion is that prices favor the "stronger" or "bigger" partner at the expense of the other. For instance, when we are told that such-and-such company "controls" a very high share of a market, many people conclude that the business in question can impose whatever products or prices it chooses on the consumer. Economists would

point to the fact that apparently powerful corporations are, in fact, perpetually threatened by consumer choice. That is why the corporate giants of
yesteryear, those industrial Leviathans that were said to "dominate" a particular market and impose their will, in many cases simply sank without a
trace. But the power of consumers only exists in the aggregate, which makes
it invisible to individual customers.

Are those folk-economic beliefs all wrong and misguided? That is the
way many economists would describe them. But even if we accepted that
the beliefs do reflect serious misunderstandings of the way a complex economy works, and if we assumed that economic theory by contrast provides a
valid explanation—which of course is really uncertain—that would not explain why these beliefs are so widespread, and why they are so important in
many people's political choices. After all, people may be wrong or misguided in many different ways, so that we should expect them to have
random views of the economy, which is not the case.[74]

An obvious explanation, it would seem, is that people take their views
on the economy from media sources, or from propaganda by political entrepreneurs. This is of course partly true, to the extent that people often justify
their folk-economic opinions by citing external sources, and that people
who have more, and more detailed, economic beliefs also have more interest
in acquiring information from media or political parties. That, however, is
only the starting point of an explanation. Even if most of people's information about the economy comes from such sources, we still have to explain
why that information grabs their attention, is stored and retrieved, as opposed to other information that is just discarded or forgotten.

Another common interpretation is that people have beliefs that fit
their interests. This is also obviously true and clearly insufficient. True, the
strongest opponents of international trade are, understandably, to be found
among those whose jobs or businesses may be threatened by foreign competition. But, as the economist Bryan Caplan demonstrated, self-interest
explanations are in fact very limited. On many issues, people's views stem
from particular biases, for example, that policies should "create" jobs or
that the creation of value is a zero-sum game, that are not clearly related to
their own interests.[75]

So, where do these beliefs come from? To explain them, we must first remember that they are not a theory of the economy. That is, people's economic beliefs consist in reflective beliefs, conscious elaborations or comments on our intuitions. This means that the beliefs are activated when relevant, but they do not necessarily come with clear descriptions of what would make them true or false. This also implies that they are contextually relevant, so that they are simply not activated in contexts where they would not produce any further inference. For example, people may sometimes think that "wealth is zero-sum" when they think of extreme contrasts in income. They may on the other hand not activate that representation when dealing with their butcher and their baker. When entertaining reflective beliefs, our minds are simply not in the business of systematically testing hypotheses—as we have seen in other domains in previous chapters. But this raises the question of why these particular views of the economy are grabbing and compelling.

Markets versus the Social Exchange Template

The economist Paul Rubin coined the term "emporiophobia" for the distrust or fear of markets in general.[76] There are many manifestations of this intuitive dislike of markets in modern societies. For instance, many people judge that market processes in the economy are probably the cause of inequality, or of inefficiencies observed in modern economy. Another, different manifestation of emporiophobia may be involved in the moral resistance to establishing a market for, say, organ transplants or children put up for adoption. Emporiophobia is very common in political discourse. How can we explain it? Perhaps our template of social exchange affords a plausible, though speculative, explanation.

Market transactions are said to be impersonal. People in modern conditions do not need to have much information about their exchange partners beyond their position (seller, buyer), the particular type of goods they sell or buy, and their posted price. There is no expectation that considerations other than price and utility will govern decisions. You can ignore the character of the baker and the butcher when you decide to patronize their shops. Importantly, there is no expectation of repeated exchange. After you have acquired the bread and sausages and tendered the agreed payment, it is perfectly fine if you never enter into any further interaction with these providers.

These features, which all constitute advantages of market transactions from an economic standpoint, may be interpreted by intuitive inference systems in a different manner. As I mentioned before, our template for social exchange requires that you track relevant information about people that goes beyond their offer and the value of the goods transferred. In particular, the template includes a strong preference for interaction with identified partners, as that allows each party in the transaction to seek and receive information about past transactions. Equally important, we have a strong preference for repeated interaction with privileged partners, because this provides some assurance against exploitation. To the extent that neither partner would benefit from breaking off the chain of transactions, we can expect that neither of them will try to exploit the other, for instance through price gouging.

Given these preferences, the special, remarkable characteristics of mass-market conditions can be interpreted by our mental systems as so many threat signals. Just as a slithering shape in the grass indicates the possible presence of a snake, the notion of one-shot interactions with unknown partners might convey the potential threat of exploitation. That is why most prehistoric trade probably followed chains of transmission between known individuals. That is also why, even in modern mass-market conditions, many people try to reestablish patterns of interaction that more closely follow the social exchange template. Most people prefer known contractors to perfect strangers. Corporate executives prefer to engage in repeated trade with known suppliers or buyers. And in some places there is an established cultural norm for such relations. This is the case, for instance, with the Chinese notion of *guanxi,* the informal exchange of favors that many people feel should provide the context for market transactions. This can be interpreted cynically as a recipe for corruption, but in more traditional environments it mostly refers to the need for repeated exchange between favored partners with ample information about each other.[77] In other cultures, there is no such explicit norm, but the tacit expectations are very much the same.

The activation of our intuitive expectations, our social exchange template, may also explain another important theme of folk-economic beliefs, a general distrust and moral condemnation of the profit motive. That, too, is a recurrent theme of political discourse, where market competition is often described as based on greed. As a result, many people, for instance, expect nonprofit organizations to be better than commercial organizations at providing people with what they need. In some studies, when prices are identical, people would prefer to buy from firms that make lower profit margins rather than from more successful ones. It would seem that Adam Smith's famous statement that the best guarantee for appropriate provision of bread and meat is the baker's and the butcher's self-interest fell on deaf ears.

Perhaps this belief in self-interest as a threat to the general interest is also rooted in our social exchange preferences and the conditions in which they evolved. One characteristic of human cognition is sophisticated mind reading, the capacity to form representations of other agents' representations and intentions. The system is crucial to human coordination and co-operation.[78] We cannot have efficient cooperation or trade without a rich description of the behaviors of other people—but also, crucially, of their beliefs and intentions. The detection of free riders, for instance, includes an automatic search for information about their possible intentions. In the context of small-scale economic interactions, given the high costs of information, detecting other people's intentions is the major way in which we can estimate the possible future benefits of transactions. People who are trying to trick you do not have your interests at heart. You should turn down their offers, however tempting they may seem. That is because you suspect that they will want to exploit you in the long run. Rich gifts wax poor when givers prove unkind, as Ophelia would have put it. Conversely, people motivated to help you may not be offering a very tempting deal this time, for reasons outside their control. You should not spurn them, however, since next time round they may be able to offer a much better deal, which they will do if they can.[79]

So, before mass markets it was highly efficient to infer outcomes from intentions. That is, the terms of exchange were much more likely to be beneficial to us in the long run, if the partner appeared to be motivated mostly by the desire to establish durable, mutually advantageous interaction with us. By contrast, if the only information available was that the partner was trying to maximize his own benefits, that would signal the kind of interaction we should avoid. Also, those who tried to cheat you in trade would be undesirable partners in hunting, warfare, and other forms of collective action outside economic exchange. That was all the more reason for tracking people's motives. In such conditions, the modern, market-oriented strategy that consists in considering what people are offering, without ever wondering why they are offering it, would be disastrous. So ancestral conditions would make it a sound strategy to believe, at least tacitly, in some form of moral contagion, whereby morally reprehensible intentions (such as the desire to maximize one's own profit and only one's own) seem to contaminate the transactions—so that we expect generalized self-interest to produce bad outcomes.

Where Conceptions of Justice Come From

In modern democracies, almost everyone agrees that justice should be the organizing principle of society, with stark differences in what people think this requirement implies. People's different conceptions of justice, at first sight, stem from adherence to specific ideologies or systems of values, so that people who value rights and rules favor equality of opportunity over equality of results, those who value effort espouse meritocratic views of society, and those who value equality as such favor redistribution of wealth. But when we try to explain why some beliefs or preferences are widespread, it is of course unsatisfactory to stop at describing them in terms of divergent ideologies. What we call ideologies are themselves bundles of beliefs that happen to be popular in some groups. If we say that people are egalitarian because they are influenced by the egalitarian ideology that surrounds them, we then have to explain why there is such a thing around them, which means explaining why egalitarianism is in some conditions attractive to some minds.

Perhaps there is another explanation. Our psychology seems to include a very specific set of intuitive inference systems to deal with exchange—and perhaps these systems are highly influential in shaping our explicit theories of justice, as well as our adherence to different conceptions of what makes a society just. As I described above, the psychology of exchange consists of distinct systems with different tasks, for example, to produce definite intuitions about who owns what, to judge whether an allocation of goods is fair, to recognize reliable exchange partners, and the like. Now it could be that, in the context of modern societies, people try to extend the scope of these intuitions to make sense of mass-market societies, and that this shapes their notions of what is just. There are, it would seem, two main paths from our intuitive exchange psychology to a conception of mass-market society.

One such path is traced by extending our intuitions about ownership and fair exchange to the vast numbers of transactions of a modern economy. Humans have precise intuitions about what is legitimate and what is fair in exchange. We find it legitimate, for instance, that people be entitled to the fruit of their work. Even children think that someone who has made a snake out of Play-Doh is entitled to keep the object, as much as or more so than the original owner of the material. We also find it fair that people sell their goods at the price their buyers will accept. We intuitively judge that transactions should be voluntary. That the baker has not sold anything today should not compel you to buy from her.[80] If we extend these intuitions, we describe mass society and its complex economy as an aggregation of such interactions, each of which is just or not, on the basis of those intuitive fairness and ownership criteria. People are entitled to the goods they possess or extract from nature, and they are also entitled to what they acquire through voluntary exchange with other free agents.

In scholarly terms, this kind of intuition underpins a conception of justice, inspired in part by John Locke, that is represented in modern theory by Robert Nozick, for instance.[81] This view of justice makes no prediction about the kinds of income or wealth distributions that could result from the accumulation of myriad just transactions. The resulting allocation might be very equal or very unequal. Some spectacular differences, for example, famous actors earning thousands of times more than nurses, should be considered just, as long as no one was coerced (in this case to buy tickets to the actors' performances). True, successful actors probably benefited from all sorts of preexisting social institutions, like the existence of theaters, drama schools, and so forth. But those advantages were also extended to many others who did not succeed in pleasing the crowds, so the difference in remuneration cannot be treated as an unjust social outcome.

The second path to explicit criteria for social justice starts from our evolved intuitions about collective action and redistribution, projecting them onto the much larger canvas of a mass society. In this perspective, the economy or society as a whole is construed as a gigantic collective action, to which everyone contributes in one way or other, and from which they may receive rewards, in the same way as a group engages in some collective task like digging a well or clearing a forest to settle a village, and everyone benefits. Naturally, the complexity of mass-market economies is such that this overall collective action comprises myriad different particular contracts and transactions, and no one at any point needs to know in what manner they contribute to the general effect. In this view, the economy as a whole produces an enormous amount of wealth, and the main question of politics is how to allocate that bounty to different individuals.

Our collective action capacities comprise intuitions about distributions. As I explained before, we intuitively recognize that people should be rewarded in proportion to their contribution. In principle, this might justify all kinds of allocations, very equal or very unequal. Additional intuitions modulate this meritocratic assumption. For one thing, humans do not generally believe that any individual's contribution could possibly be hundreds or thousands of times greater than anyone else's. It seems difficult to believe, for instance, that famous and highly paid actors really contribute to society something that is thousands of times more valuable than their poorly paid competitors. Also, we intuitively favor more egalitarian sharing when we cannot easily trace the connection between effort and result. That is the case in hunting, where similar effort can have very different results, as opposed to the more predictable returns of food gathering. Finally, our intuitions evolved in the context of small groups, in which it would be imprudent to leave some individuals with no share of the proceeds at all. Sharing of unpredictable resources is intuitively perceived as a form of insurance, so that we cannot altogether alienate those who did not contribute much at a particular time.

Taken together, these intuitive expectations and preferences, applied to the inscrutable complexity of a mass-market economy, lead people to focus on resulting allocations as what is just or unjust. They also suggest that extremely unequal distributions cannot really be just and, most important, that it may be morally justified to correct them. This perspective, based on entrenched intuitions about collective action, does not by itself explain how such a correction could be made. But it makes some policies, like progressive taxation or limits on income, appear intuitively just.

This perspective is made coherent and explicit in John Rawls's theory of justice.[82] Rawls stipulated that social orders would be just if they provided essential liberty (for example, to hold personal property, to engage in transactions) as well as equality of opportunity. Further, Rawls's "difference principle" stipulates that, as wealth increases, it should be distributed in a way that increases the resources of the least advantaged (although it may also benefit better-off individuals).

These two visions underpin the two main kinds of theories of justice, focused either on processes—transactions must be just, and the resulting distributions will be just—or outcomes—distributions must be fair, and measures that make this more likely will be just.[83] The theoretical elaborations of these principles of course go much further in their technical sophistication.[84] We do not need to explore them, as our problem here is not to advocate between visions of society but to understand how they could become compelling and natural to human beings. What seems to happen, in both perspectives, is that an abstract conception of justice attracts attention, is made cognitively salient, because some of the information it contains happens to fit some of our intuitive systems.

People do not just reflect on these issues, they also feel that something should be done, they are outraged at what happens, they try to bring others around to particular views. The activation of intuitive systems makes sense of these emotions and motivations. The cognitive systems that guide fairness in exchange, or govern ownership, or monitor the distribution of goods from collective action are precisely designed to trigger emotion and motivate behavior—otherwise they would have been of no evolutionary advantage. Ownership intuitions result in a vigorous defense of what we extracted from the environment, and a robust motivation to help others guard what they extracted against intruders. Our free-rider detection system delivers a powerful desire to curb the activities of cheaters, reduce their gains, and advertise their misdeeds. Once such systems are activated in the description of large-scale social phenomena, they trigger the same emotions and motivations.

Prosperity and the Paradox of Justice

For most of prehistory and history, people lived at a subsistence level that more or less corresponds to the production level among today's poorest in the poorest countries, estimated at about $2 to $3 a day. Then, starting with the industrial revolution, prosperity rose at an ever-increasing rate, producing a "hockey stick" curve in wealth in most human societies. As Matt Ridley has argued, it is very difficult for us to appreciate how much our conditions have changed compared to the conditions experienced, in fact endured, by most members of the species until this occurred. People live longer than before, can be cured of a large number of conditions that would have killed them a few decades ago, and can eat much better food, drink cleaner water, and spend their time in well-heated shelters, as well as communicate with millions of other individuals and tap knowledge accumulated for centuries, all at a much lower cost than before.[85] And the most tremendous change in the world economy over the past fifty years has been the immense reduction in poverty, the world over. As I write this, it is estimated that "only" one billion people still live at the Malthusian level of $2 a day or little above that. The quotation marks signal that it is a scandal that so many are still in poverty, but this decrease is an unprecedented, and to most people largely unexpected, development in human history. Nothing of that magnitude had ever happened before.[86]

For decades, economic historians have been arguing about the particular conditions that launched the industrial revolution in the eighteenth century, the special factors that had made England, Holland, and northern Italy mercantile hubs before that, and what made England in particular such a center of technological innovation. Though economists disagree on the "why then" and "why there" questions, they all agree on the "how" question, that is, How did general prosperity encompass the entire world? Here the answer is that prosperity rises, first slowly and then increasingly fast, in all places where people can engage in peaceful and voluntary exchanges. Trade and the innovations that it makes possible provide the only known escape route from poverty.[87]

This creates a paradox. Trade did not have that effect until it reached a particular degree of expansion, involving millions rather than thousands of individuals, thereby involving these agents in webs of interaction that no human mind can track. All we can observe are emergent effects. But the tools we have in order to make sense of these aggregate effects were designed to explain small-scale interactions. The complex coordination of market economies, in which millions of individuals cooperate without any plan, is beyond the scope of our psychology of exchange. As a result, a great deal of production is tacitly considered, at least by part of our cognitive system, as a given, as part of the landscape, so to speak. That is, we have cognitive systems that represent the existence of these goods and services, but without representing the reason why they appear. A fortiori, we do not represent, at least not in intuitive terms, the fact that other goods and services might have appeared, or that the ones observed might never have been produced.

If a great deal of production is just given—to be more precise: if some of our cognitive systems take it as a given—then it is bound to activate the intuitive systems concerned with communal sharing. That is because a good part of the available goods and services are mentally processed as a windfall, as something that is clearly there, but whose origin is simply not represented. If you and your hiking companions find a large banknote on a mountain path, this may activate questions and intuitions about sharing (for example, Do all the hikers get a share of the find? Does whoever spotted it first get a larger share?), but these are totally independent of questions of origin (Who left it there, why?). I am not suggesting that anyone explicitly represents industrial and commercial production as a windfall. All that is required is that some mental systems may handle information about production without ever searching for information about its origin.

If all this is valid, our conceptions of justice seem to lead to a paradox. The reason humans could develop trade, and expand it far beyond the confines of small-scale production and local consumption, is that we have a set of evolved dispositions for mutually advantageous transactions, based on strong intuitions and motivations concerning ownership and participation in collective action. Because of these mental dispositions, we created an extraordinarily complex economic world, and the prosperity that comes from this complexity. The world created consists in countless products and services, whose existence cannot be explained by our intuitive systems. They seem to appear, but no intuitive system represents the conditions under which they appear. So they are treated by some mental systems as a windfall. This in turn activates our communal sharing preferences and intuitions, which make certain conceptions of justice, notably the distribution of available wealth, both intuitive and compelling, that is, easy to process and convincing. But the notion of redistributing wealth violates some intuitive expectations, to do with effort and reward—those who contribute more should receive more—and of course ownership—those who produce are entitled to what they produced. Redistribution implies some limits to these expectations. Some people may contribute a lot more than others but receive only a little more than others. Some may have to relinquish part of what they produced, in the form of progressive taxation. So the policies intuitively preferred because of one intuitive system (sharing) clash with preferences from another intuitive system.

There are of course many sophisticated ways of going past this conflict between different sets of intuitive preferences. But that is the point—they are sophisticated, they require the work of scholars, and it takes some effort to learn them, because our mental equipment does not provide us with an intuitive resolution of this inconsistency. Humans seem to generate trade because of fairness, and trade creates results in so much impersonal production that the imperatives of fairness seem to clash with the requirements for trade.

Can Human Minds Understand Societies?

Coordination, Folk Sociology, and Natural Politics

HUMANS WERE DESIGNED BY EVOLUTION to live in societies, but they may not understand how societies work. This may seem paradoxical. Man was classically described as the political animal; many people in many places seem to be attentive to political processes and be emotionally engaged in political action; and many people, it seems, even enjoy talking about politics. Political programs, political disputes, and political arguments, not to mention revolutions and reform, all convey general ideas about the way a society works and ought to work, how institutions are created and maintained, how different groups and classes interact, and so forth. Such ideas are not the preserve of specialists; they fill everyday debates and justify opinion among all or most citizens of mass societies. But are these notions accurate?

Optimistically, we might think that our commonsense political theories must get most things right, otherwise we could not have complex societies, we would not know how to behave in such societies. But that is not a very sound argument. One can behave in an efficient manner in a market economy yet have misguided views of what makes such an economy work. In the same way, it is possible that our minds were shaped, through human evolution, to make societies possible, even the large-scale ones we now live in, but not shaped in such a way that we actually understand what makes societies hang together.

Theories about the way society works are pervasive in modern politics, but they are much older than that. In all human societies, people have some notion of what social groups are, how they are formed, what political power consists of, why some individuals have higher status than others, and so forth. The anthropologist Lawrence Hirschfeld coined the term "folk sociology" to describe spontaneous understandings of social categories and groups.[1] I shall borrow this term and expand it to cover people's notions of how society works, what its components and their interrelations are. So the question is, How do people build their folk sociology? And of course, Is our folk sociology accurate? The question is not just academic. Large-scale societies rely on deliberative democracy for their collective decision making. Deliberation requires, at least in its ideal form, the rational comparison of policies based on an accurate understanding of what happens in society. But is it possible for human minds to reach such an understanding?

Transparent Deliberation: The Republic of Te

As a starting point, we can consider the case of a small-scale polity that illustrates what could be called a deliberative ideal. The community of Te, in the Mustang region of western Nepal, is organized as a small republic. The term is perhaps misleading. Te is somewhat autonomous, but the village is also part of a federation of five communities, itself included in the former kingdom of Baragaon, which for centuries has been subsumed into the Nepali kingdom. This latter's authority manifests itself by levying taxes, and very little else. The citizens of Te themselves manage most of their collective affairs. Like most denizens of the high Himalayas, they depend for their subsistence on their cattle and barley, and they are part of extensive trade networks that follow the valleys. Te itself is an extraordinarily compact aggregate of about a hundred houses, some on top of the others, huddled together on a small patch of terrain to leave room for the grazing meadows and the small fields irrigated by a complex system of aqueducts.[2]

In Te, citizens not only participate in collective decision making but also periodically rewrite their constitution—the document that enumerates public offices, their tenure and associated duties, and the mode of election to such offices. The constitution also specifies rules for collective action, such as the maintenance of irrigation canals, aqueducts, and grazing commons, the organization of periodic ceremonies and sacrifices to local spirits, the fines levied for wandering cattle, and other such matters of common concern. Every twelve years, an assembly of senior men from each household adjusts the rules or introduces new ones. Participation is compulsory, on pain of heavy fines. After two weeks of constitutional debate accompanied by votes on each proposal, a consensus set of rules is agreed on and recorded on paper by a Buddhist monk, to serve as a document of reference in case of subsequent disputes.

There are several public offices in Te—all of them are of course part-time occupations. These include the offices of three headmen, in charge of day-to-day decision making and of representing the community in relations with outsiders, as well as four constables to monitor compliance with the rules, for instance concerning the use of grazing commons. All these officers are chosen by drawing lots, once a year. They take an oath of service and work under the close supervision of all the other citizens.

Politics is necessary in Te, because no one could survive the very harsh conditions of the Himalayan ecology without engaging in extensive collective action. There are grazing commons for the cattle, the use of which must be regulated. The small fields require controlled irrigation, and the aqueducts and tunnels need periodic repair. Te must pay taxes to its political masters, once in Baragaon and now in Nepal. All this requires some measure of concerted, carefully monitored communal cooperation. That is why the rules list many specific fines for taking advantage of collective goods without paying the required price, for instance by appropriating firewood from communal fields. Also, one cannot just shirk one's obligations by making oneself scarce when the going gets rough. The constitution stipulates that "tax exiles" are not allowed to come back to Te or even trade with its citizens.[3]

Te is a place where politics is transparent. The rules that make up the constitution are entirely clear to all citizens. They specify such matters as "the constables shall monitor the fields twice a day" (to deter free riding on the commons), "constables who leave the village for more than a day shall pay a fine of 100 rupees," and "people will not remain in mourning [thereby avoiding communal work] for more than forty-nine days."[4] There is no ambiguity or uncertainty about what these regulations mean and how they will impact the welfare of each individual or constrain his behavior, and therefore why each agent would be in favor of or against a specific proposal. The process, too, is transparent, as the citizens themselves are involved in all decision making. Finally, decisions are reached via deliberation. Each participant can make a case for rules that would advantage him, and each has to negotiate with everyone else to reach a consensus. All this constitutes the deliberative ideal, a very natural way, for many people, to think about the political process. However distant from more familiar mass societies, life in Te serves as an illustration of what we intuitively expect from politics.

Social Complexity and the Origins of Politics

Politics is not as transparent in the mass societies in which most of us live. To understand how our minds handle large-scale decision making, we must first step back and outline the way societies could grow in the first place. Human beings evolved in small-scale foraging groups. But they also managed to create large-scale polities, through a process that led from bands to larger tribes, chiefdoms, city-states, kingdoms, and empires.[5] So a fundamental question for anthropologists is to explain how this scaling-up process is possible, how large-scale chiefdoms emerged from tribes and how impersonal state institutions emerged from personal tribal authority.

Most of human evolution took place in the context of small bands of nomadic foragers, probably numbering twenty to forty individuals, moving as a group to optimize the return from hunting grounds and seasonal plants. These bands were almost certainly included in larger tribes, with a common language, occasional meetings and marriage exchanges between bands. Extrapolating from what we know of present nomadic foragers, and using the archaeological record as a guide, it would seem that when resources were limited, these remained mostly small-scale, egalitarian societies. By contrast, foragers created social ranks and hierarchies in places where a more generous environment allowed some accumulation of wealth. For instance, native societies of the American Northwest developed social ranks and complex political systems.[6] The size and complexity of human societies changed dramatically with the appearance of agriculture, with its development of sedentary groups and a much higher population density. Many groups specialized in horticulture, a small-scale extraction of resources from a few staple crops, combined with extensive hunting and foraging. Agriculture and pastoralism ushered in classical peasant societies, where the surplus from an agrarian economy supported a class of aristocrats and the institution of kingship.

Early "evolutionist" anthropologists tended to see these trends as uniform and irreversible. But the transition from nomadic foraging to populous agricultural centers proceeded by fits and starts, with quite a few reversals when ecological conditions made agriculture unsustainable. Not all large chiefdoms became states, and some city-states dissolved into the surrounding tribal societies. Also, state formation and the relationships between emergent kingdoms and surrounding tribes took on very different forms in different places.[7]

The emergence of large-scale societies raises the crucial question of mental dispositions and preferences. What kind of psychology is required to build and maintain large and complex polities? How is the mind equipped for that? Humans had to have whatever it takes to create complex political systems and solve the many coordination and cooperation problems that result from assembling large numbers of people in a single polity.

One should of course consider the possibility that the populations that engaged in agriculture, animal husbandry, and urban dwelling were somehow genetically different from populations of foragers. Indeed, with increased knowledge about ancient DNA, it seems that there appeared clear differences between those populations. There is strong evidence for rapid and profound evolution by natural selection over the Holocene epoch, and for fast differentiation between populations. But these differences are consequences rather than causes of the shift to a Neolithic lifestyle. Humans who lived in crowded quarters had to survive larger exposure to more pathogens, and their immune systems changed in response to that evolutionary pressure. The same goes for exposure to the bacteria and viruses brought by pigs, cows, dogs, and other domesticated animals.[8] The radically new diet, based on high-calorie staple crops like millet and barley, also contributed to profound genetic changes. Finally, dependence on milk products created a selective pressure for lactose tolerance in some populations in parts of northern Europe and West Africa.[9]

So, in all probability, the evolutionary equipment that made it possible to create large-scale societies in the first place was largely present before the Neolithic Age. But what is that equipment? It must consist in capacities and dispositions that allow human groups to scale up. The popular cliché that ours are Stone Age minds is mostly accurate, as our complex mental dispositions required long evolutionary time to emerge as a result of variation and selective pressures. But that ancient mind clearly allowed the emergence of the institutions of complex societies.[10]

Scaling-Up Toolkit (I): Collective Action

Assembling a large-scale community does not reduce to doing the same as in small groups, except with more people. The reason is that a large grouping creates new problems. For one thing, it requires coordination with strangers, in contrast to small groups, where most everyday activities involve partners who know each other, and many of whom are kin. So it must be possible for humans to join others in the pursuit of some common goal that is beneficial to all, beyond the simple imperatives of cooperation with those who share your genes. Also, it must be possible to get all those individuals to do the right thing at the right time. Any complex society is a marvel of coordination, as thousands or millions of individuals behave in the appropriate way given what others do, like a gigantic orchestra. In other words, humans would need capacities for collective action, and for the productive combination of different agents' contributions.

In chapter 1, I described coalitional psychology, the set of capacities and motivations that make it very easy for humans to build and sustain large and cohesive alliances, producing the appropriate intuitions for rivalry between high-solidarity groups. Now the capacity to create coalitions is only one among the many consequences of the human dispositions and motivations for collective action. Coalitions are simply a form of collective action applied to rivalry between alliances, but humans create collective action in many other ways and in many other contexts.

Complex interaction is clear from the time of *Homo erectus*, at least half a million years ago.[11] Early humans engaged in collective hunting and cooperative parenting. Both constitute collective actions the mutual benefits of which must be estimated in advance, and therefore require a bet on other participants' goodwill. Archaic humans could run animals to death, gather them in a battue, and drive them over cliffs—all operations that require some degree of efficient coordination between different agents.[12] In these kinds of cooperative ventures, monitoring of other agents' participation is not too difficult, and free riders can be punished by withdrawing cooperation. In more modern contexts, people build and maintain institutions that broaden the scope of collective action to include millions of citizens in joint projects. People pay taxes and serve in the military. But they also build large corporations and join associations or political parties. So it is crucial to understand how and why humans manage to create and sustain collective action.

Classical social science has not always been very successful at explaining that. Rousseau famously described early man as enjoying a carefree independent existence, until the development of agriculture and industry ruined it all by making some individuals depend on others.[13] One could hardly be more wrong. Indeed, Rousseau should have taken inspiration, had he known about him, from an earlier and more empirically minded social scientist, the Tunisian Ibn Khaldūn, who emphasized the necessity of *'assabiyah* or group solidarity at the most primitive stages of social evolution.[14] We now know, from the archaeological record, that interdependence and cooperation are hallmarks of human existence from the earliest times in human prehistory. It is a characteristic of our species. It makes sense to wonder how natural selection resulted in our motivations and capacities for collective action, and how these are engaged in actual social interaction.

For a long time, this has been a question addressed mostly in economic theory, and the answer was somewhat paradoxical, and certainly frustrating—most economic models suggested that collective action was very unlikely. In fact, if all agents followed the imperatives of utility maximization, they would never even try to set up such mutually beneficial forms of cooperation. There are of course many variants of this proposition, and many nuances that I shall not explore here, because the line of reasoning is simple. The problem of collective action is, obviously, the potential for free riding. If we the people all agree to overthrow the dictator, we can accomplish the overthrow without too much cost (even the dictator's police will side with us!) and we all will benefit. But you as an individual will benefit even more if you let the others do the job. The best strategy is not to show up before the evening of the revolution, and to be as late as possible.[15]

Naturally, economic theorists are aware that collective action does happen, indeed happens all the time in all human societies. But they are happy to describe this as somewhat irrational, as a deviation from normative theory that can be explained by extraneous factors, like people's unjustified confidence ("Surely the police will join us in overthrowing the tyrant!"), their exaggerated beliefs in other people's trustworthiness ("Surely all those who pledged to join the cause will do just that!"), or perhaps in terms of intermediary benefits that could be gained by participating (there may be prestige and material benefits in being among the revolutionaries).

But there is something awkward in a theory that describes as odd or deviant what humans do spontaneously and so often. The fact that humans everywhere engage in collective actions in many different domains, and in all known human groups, would suggest that classical economic models were perhaps based on the wrong assumptions. Indeed, formal models of collective action generally assume that participants care only about the end goals, not about how they are achieved, do not care who the other participants are, as long as they join, and are interested in their own payoff, that is, how the collective action will profit them, and indifferent to how it will profit others.[16] But collective action as practiced by most human beings, in most known societies, demonstrates exactly the opposite. Participants in joint hunting or collective parenting, or modern institutions for that matter, are intensely interested in the ways in which the collective goal is achieved; they very much care who else is involved and how much they can be trusted, and they monitor attentively the distribution of benefits from collective action. In formal, game-theoretic terms, human collective action is not an extended Prisoner's Dilemma, where each participant knows that the others would benefit from defecting, regardless of whether she cooperated—so that everyone is stuck with the strategy of defection. Collective action is more similar to what Thomas Schelling described as a tipping game, where payoffs depend on the number of people who will join—and therefore information about other agents' preferences is crucial.[17]

Collective action is indeed difficult, in theory, for many reasons. First, in most cases collective action requires that we defer gratification. Foragers who join a hunting expedition gain nothing at the outset. The potential benefits are all in the future, while the danger or effort is very much in the present. That is true also of barn raising, long-distance trade, or investment in joint-stock corporations. But that did not stop humans from engaging in such operations, perhaps because imagining possible benefits helped them counter the natural tendency to discount the future.[18] Second, the economists were right to point out that defection is possible and, if widespread, would jeopardize any collective endeavor. But in actual human societies, communication between agents minimizes that danger, because it makes it possible to impose large costs, in terms of reputation, on those who wish to defect when the going gets rough. As I explained in a previous chapter, the possibility of partner choice and the reality of efficient communication alleviate the problem of defection, as they make defection a losing strategy in the long term, especially in the small-scale communities we evolved in, where we shall all meet again, and therefore we cannot afford to alienate all of the people all of the time. To sum up, the theoretical problems of collective action are real, but natural selection seems to have solved them.

Scaling-Up Toolkit (II): Hierarchies

Collective action would not be efficient, and certainly could not scale up to include thousands or millions of agents, without the organizing glue provided by the human capacity to build hierarchies. Most human interaction requires not just that different agents do different things but also that decision making be allocated to particular individuals, rather than remain diffuse in the group. This is obviously true in large-scale societies. We are all familiar with work hierarchies, with the fact that organizations like schools, corporations, armies, and political parties comprise different ranks with different authority. There are also hierarchies on a smaller scale. When the entire village shows up to erect a barn, organize a feast, repair a dam, or organize a religious ceremony, there is invariably some degree of division between decision and execution.

What is the origin of these hierarchies? There are dominance hierarchies in many species, specifically in most primate species close to humans. Groups of chimpanzees, for instance, have dominant males and females. In many species hierarchies are visible in terms of access to resources, notably to mating. Dominant males in species with high intrasexual competition have privileged access to females. Access to other resources can also be modulated by rank, as in the hierarchies of females in some primate species. These are dominance hierarchies, whose main effect is a specific allocation or distribution of available goods. So it would be tempting to assume that humans simply preserved old dispositions for hierarchical differences, with consequences for both reproduction and, in modern human polities, practical control of other individuals' actions. In this view, political dominance in human groups would just be another version of the dominance hierarchies of our common ancestors among closely related primates.[19]

But this may be a misleading comparison. Human hierarchies are very different from the dominance orders found in many other species. First, as the economist Paul Rubin pointed out, human hierarchies are not (or not just) relevant to the distribution of goods. They are also (and mostly) production hierarchies, that is, ways of orchestrating different individual's contributions to a task. Humans readily build hierarchies in almost any domain where the actions of more than one or a few individuals are required. And each single agent can be part of multiple, independent production hierarchies.[20]

The starting point of production hierarchies is the fact that different agents may not have the same information or skills. This can be demonstrated in purely formal game-theoretic terms, even for activities that involve only two players. If one player is slightly better or faster at taking particular decisions, it may be in the interest of the other player to free ride on these skills, as it were, by following rather than leading, thereby subcontracting the decision making to the more competent agent.[21] Such dyads are common in human interaction, and they can of course expand to include several followers of one leader. The evolutionary psychologist Mark van Vugt speculates that such coordination needs may have provided the context for the emergence of leadership and followership strategies in human minds. Specifically, the evolutionary conjecture is that human minds can detect situations that require coordination for complex action. They can also measure the costs and benefits of leader and follower strategies, given a particular interaction, adjust their behavior appropriately, and signal the role they intend to take in that interaction to other agents. There is indeed evidence that, even faced with an entirely novel task and previously unknown partners, humans quickly identify potential hierarchical organization, and that members of a team usually agree on the most efficient rankings.[22]

The Will to Power versus Human Anarchism

Hierarchies in human groups are not just about production—they often congeal, so to speak, as a system in which some individuals are in power, that is, in the leader's position in hierarchies that extend to multiple domains of behavior. Why is there unequal power in (almost) all human societies? Here again, one benefit of addressing the question from an evolutionary standpoint is that it forces us to reconsider very familiar phenomena as oddities that we should try to explain.

For one thing, why would individuals be interested in gaining power over others? The answer is more complicated than it may first seem. Political systems depend on the availability of individuals prepared to pay some cost, sometimes considerable, to occupy particular positions in coordinating hierarchies. We know that whatever the office, there will be individuals prepared to occupy it. A simplistic explanation would be that power positions in many cases confer direct benefits—the example of the warlord or autocratic king comes to mind as a prefect illustration. Modern history has shown, however, that the supply of candidates remains just as abundant when the perquisites of office are much more limited. This suggests that there is an intrinsic motivation for power in many human beings, as many classical philosophers suggested, a naked "lust for power" in Nietzsche's words.[23] But why would that be the case?

Evolutionary psychologists have long noted that, in most societies, political power has important consequences for men's mating prospects, and to a certain extent for women's fitness. Women may gain from their hierarchical position by having a large network of allies who can provide help and support in case of need. For males, the reproductive advantages of high status are more direct. Men of higher rank have more access to the most desired women, and they have the resources to bring up stronger and more numerous offspring.[24] Also, as many small-scale societies in human prehistory engaged in some measure of rivalry or warfare between groups, leadership in combat provided more evidence of one's capacity to recruit social support, and to protect one's kin. This correlation between power and fitness is so clear, and has been present through so many historical and economic changes, that it certainly shaped our psychological dispositions. Indeed, in terms of mating criteria, we know that cues of rank and influence are among the most desired traits in males.[25] This would explain why the motivation to acquire political power is not narrowly opportunistic—it is not a function of the actual or expected benefits of political office, although those are of course important stimulants as well.

Power in this evolutionary perspective is one of the many proxies for fitness that direct our behavior—that is, observable goals which, over aeons of evolutionary history, coincided on average with higher fitness. Many of us, in many different situations, feel compelled to seek positions of dominance when possible because that seems intrinsically desirable (that is the proximate explanation), but that itself makes sense because relative dominance brought about higher fitness (that is the ultimate explanation), on average, in ancestral environments.

But human preferences also provide a buffer against this lust for power. As the anthropologist Christopher Boehm pointed out, there is very little concentration of power in small-scale societies. To the extent that there are leaders, these are generally legitimized by their experience and proven competence. But people are quick to deflate the claims of leaders desirous to broaden their influence or bully their peers.[26] Anthropologists have documented this urge to circumvent would-be despots in many small-scale societies. Indeed, in many places being a chief is a mixed blessing, as one is supposed to resolve conflicts, to supply resources for collective ceremonies, to provide support when needed, and generally to redistribute resources all around, on pain of being demoted or expelled.[27]

A crucial factor here is that humans evolved in nomadic foraging groups, where people could at least in principle vote with their feet, and therefore had a not negligible bargaining advantage over autocratic chiefs. That may be why the motivation to resist overreaching political domination has been observed in all human societies. That would also explain why, in agrarian societies, this desire to limit the accumulation of political power could not lead to more than occasional acts of rebellion. Leaving one's group meant leaving the land, almost invariably a sure path to starvation. But before the advent of agriculture, in the foraging groups in which our political psychology evolved, the balance of power was not as tilted as the titles of chief or headman would suggest, and it allowed the expression of what Boehm calls an entrenched egalitarian motivation in human social life.

The term "egalitarian" may be misleading, if we take it to mean that people want everyone to have the same role in decision making. Rather, the motivation is to circumscribe hierarchical decision making so that it does not lead to political exploitation. You accept that some individual should lead the hunt, but you do not necessarily accept that this gives him a right to determine who marries whom. Humans are keenly sensitive to the risk of exploitation, as exploitative strategies are always possible in collective action. Accumulation of power is exploitation, if political power accrues fitness but does not result in some compensatory benefits for other individuals in the group. To put things in abstract terms, above a certain threshold, the accumulation of higher fitness by decision makers, which is a loss for all others, becomes high enough that it matches the cost of rebellion or defection.

So, rather than being egalitarian, the motivation that limits the ambitions of tribal chiefs could more properly be called anarchistic. The anthropological evidence from small-scale societies suggests a form of political organization that may well be typical of our species, in which some individuals indeed seek power, but other people monitor the amount of power accumulated and intuitively construe it as a potential threat. This aspect of our political psychology is congruent with our preferences for collective action, for the voluntary association of individuals in the pursuit of collectively beneficial goals. Obviously, this does not mean that humans are naturally anarchists in the modern political sense of the word. But the label may be appropriate to denote the strong human motivation, documented in the most diverse cultures, for mutual aid, voluntary associations, and the self-organized management of commons, away from overbearing despots.[28] Indeed, Peter Kropotkin, one of the theoreticians of modern anarchism, also noted this connection between what he called "mutual aid" among "savages and barbarians" and the human motivation to hold power in check.[29]

Folk Sociology

As humans started to live in large groups, beyond the confines of small foraging bands and tribes, they also developed explicit descriptions of their own societies, a folk sociology. People who live in a kingdom have mental representations of what a kingdom is like, what the rights and duties of a king are, what makes the king special, and so forth. The same of course applies, mutatis mutandis, to those who live in a city-state or in a large nation or empire. We all have some form of folk sociology. It is based on tacit assumptions about social groups, about power, about the nature of social norms—assumptions that we rarely entertain as conscious thoughts but that nonetheless govern our representations of social and political matters.

Folk-sociological understandings appear, on the surface, as diverse as the kinds of societies they aim to describe and explain. We do not expect to find the same views of society in forager bands, chiefdoms, large kingdoms, and modern states—for the simple reason that people are attempting to describe and explain different realities. But there are common principles to these varieties of folk sociology, generally assumptions that derive from the way our minds represent social mechanisms. They seem to appear in very similar forms in the most diverse kinds of societies.

Principle I: Groups Are Like Agents

One major feature of our folk sociology, found in the most diverse societies, is that we spontaneously construe human groups as agents. For instance, we talk about villages or social classes or nations as entities that want this, fear that, take decisions, fail to perceive what is happening, reward people or take revenge against them, are hostile toward other groups, and so on. Even the workings of a small social group like a committee are often described in such psychological terms: the committee realized this, regretted that, and so forth. Social groups are represented as having psychological states and processes characteristic of human minds, from perception and attention to memory and reasoning, as well as typical emotions like envy, gratitude, hatred, or friendship. In modern societies, the state and its bureaucracies are also represented as agents, in both newspapers reports and ordinary conversations, with phrases like "The government is trying to . . . ," "The Pentagon will not accept that . . . ," and so on. The metaphor of agency is widespread in our understanding of international relations, with phrases like "China will want to . . . ," "Russia is not intimidated by . . . ," and "England is reluctant to do this . . ."[30] This is not just a modern phenomenon. In many tribal societies, people talk about collections of individuals as distinct groups. Lineage societies, for instance, have distinct descent groups that are often considered to be different agents—such that one can say that the so-and-so lineage "wants this" or "resists that," and there is nothing strange in such talk. In many places in the world these days, ethnic groups or social classes play this role, and it seems self-evident that each group has specific goals or intentions.

The importance of describing social behavior in terms of agents extends to another very common feature of folk sociology: the tendency to talk about large collections of people in terms of generic agents. For instance, a debate about wages, about the consequences of having a minimum wage, are conducted in terms of what the "employees" and "employers" will do. In the same way, people will say that "women" want this or "men" do that—again, taking a generic agent as a simplified description of a specific population.

Principle II: Power Is a Force

A second, equally important element of our spontaneous folk sociology is our understanding of political power. In most human groups, one finds power differences between individuals and categories of individuals, and people have explicit notions to describe, and to some degree explain, these differences. What makes a chief? Why do we obey rulers? Surprisingly, the folk concepts of political power are not systematically studied in political science.[31]

In many places, people construe "power" as a substance or a special quality attached to some people. This is manifest in such phrases as "she has power," "she lost power," "his power increased," and so on. This is not just a Western or European way of speaking. Such metaphors are familiar from many tribal societies, chiefdoms, and early states, where the persons vested with authority are thought to posses some special, nonphysical quality that gives them, precisely, authority. In traditional Benin, for instance, in West Africa, people mentioned the king's (or nowadays the president's) possession of *acé*, an undefined quality, as the source of political authority.[32] For another, perhaps more familiar, example, consider the Polynesian notion that chiefs are in power because they are endowed with *mana,* a form of efficacy that ordinary people simply lack. The notion is found, for example, in traditional Maori polities, but was and to some degree is still widespread in Melanesia and Polynesia.[33] Hawaiian kings and chiefs, for instance, had mana and had more of it than others, and this quality or condition was what made the state prosperous. This quality could be compromised by the presence of impure commoners, which is why these had to avoid contact with the nobles and kings and their possessions, which were *kapu* ("separate," hence our word "taboo"). Indeed, only the priests and their attendants could see the king.[34] In many languages in the region, the term "mana" is commonly applied to tools that do the job, engines that actually start, crops that grow . . . and people who can exert influence on others.[35] So political power was simply construed as a force emanating from the king (and in many places, flowing from the gods) that made things work. Indeed, Hawaiians considered that relatives of chiefs and kings were less "sacred" as they were further removed from the center of power, which suggests they did consider it analogous to a physical force that decreases with distance.[36]

Sacred kings are not found only in Polynesia, obviously. Such notions were or are common in many places in Africa, Asia, and pre-Columbian America. Kings were described as essentially different from their subjects, and because of that difference they were subject to many taboos and prescriptions. Many African kings, for instance, were avoided as much as revered. The fact that they had power also meant that in many cases any contact with them was dangerous. In many instances the king's body is described as an analogue of the nation, an extension of the king's "body politic" to the entire kingdom. For instance, the Akwapim kings in Ghana are not allowed to walk unassisted, lest they may fall and precipitate the kingdom's disintegration. Because they stand for the nation as a whole, they are ousted if they fall ill or fail to produce children. In the same way, in many places in Africa people took the view that the king should be executed if he became ill or weak.[37]

So power is often described as a substance attached to specific persons, and its operation is construed as analogous to physical force. In the conventional metaphors of English, people have power and exercise power. We conceive of someone with power as able to "push" others toward certain behaviors (as a physical force can move objects); we say that people who did not follow the leader were "resisting," that they were not "swayed," they will not be "pushed around," and the like.

Principle III: Social Facts Are Things

A third important theme of folk sociology is the tacit assumption that social norms and institutions are somehow external to people's minds, that they are a social reality, a set of actual things, over and above what people think of them. This description may seem unduly metaphysical—and of course most people do not explicitly think of institutions in this abstruse manner. But they do construe norms and their effects as some kind of reality "out there," in some unspecified way. A few examples may help at this point.

In Western countries the issue of the legal status of same-sex unions has been the occasion of ardent disagreement on the way norms of the family could or should be changed. Opponents argued that same-sex couples would undermine traditional family norms, or that they would not provide a suitable environment for raising children, or that legalizing gay marriage would take society down a slippery slope to polygamy or other even more radical changes. Those in favor argued for equality of rights between different-sex and same-sex couples, based on the traditional liberal view that any behavior is permissible if it does not impinge on other people's liberty. But both sides agreed on some premises in that discussion. For one thing, they all were concurred that whether something is or is not called marriage is important, that naming matters. Some opponents, for instance, were prepared to grant same-sex couples all the legal rights and duties of married couples on the condition that such contracts be called something other than "marriage." Proponents of legal equality could not be satisfied by such measures. They, too, thought there was some actual value in using the term "marriage" as opposed to "civil union" or other legal euphemisms. Another point of tacit agreement, less obvious but equally important, was that there is somehow such a thing as marriage, independent from what we think. This is clear in such conservative statements as "marriage is the union between a man and a woman," implying that what happens in fact (heterosexual marriages) is what should happen. The statement strongly suggests that, even if we all agreed that marriage could unite two men or two women, we would not have changed marriage—we would simply be mistaken in using that label, in the same way as calling salt "sugar" would not make it sweet. And some champions of reform agreed with that assumption, but added that marriage really is the union of two individuals with a strong commitment to each other, and therefore does apply to all such pairs, heterosexual or not.[38]

Because norms and institutions are tacitly construed as an external reality, they are often justified in a circular fashion. For instance, anthropologists have often inquired why a particular ritual *x* included specific actions, only to be told that these actions were there because they had to be included, if you wanted to perform *x* rather than another ritual. Among Fang people in Cameroon, diviners sometimes perform the *ngam* operation, whereby the specialist uses a tarantula to reveal, for example, who is a sorcerer, who is responsible for someone's illness, and so forth. To reach a diagnosis, the diviner plants several twigs, or porcupine quills if available, in a circle in the sand. A tarantula is then let loose at the center of the circle, and a calabash pot covers all this. After a while, the calabash is removed and the spider escapes. Which twigs were knocked down by the spider, and in what direction, provides signs that the diviner can now interpret.[39]

Being an anthropologist, I was professionally compelled to ask the participants in this operation why one should use a spider rather than any other animal, why they chose that species of (rather dangerous) spider, why one should recite particular ritual formulas before lifting the cover, and other such details. Like many an anthropologist before me, I was dismayed when my informants, instead of providing a rich symbolic exegesis of their ritual, simply told me that they had to proceed that way in the ngam divination ritual, otherwise it would not be a ngam divination ritual. One was of course perfectly free not to perform ngam, but if ngam is what you wanted, then these particular actions were required. This way of speaking may seem circular, but it makes sense if the institution is seen as external to people's minds. My question, Why use a spider rather than a mouse? seemed strange to people, as it implied that the particular rules of performing the ngam ritual are the result of the Fang people's choices. But my interlocutors saw it differently, and they implicitly construed it as an objective fact that the Fang people had discovered—just like the fact that salt dissolves in water is an objective fact, regardless of our thoughts about it.

In many places, this description of social facts as an external reality is applied to the entire set of norms and notions particular to a group. For instance, as a result of colonization and forced cultural change, people in Melanesia use the creole word *kastom* (from the English "custom") to describe a large set of traditional values and practices, not just the rituals banned by colonizers and missionaries, but also religious ideas about ancestors and their monitoring of human existence. The term was mostly used, for decades, as a rallying cry for resistance to colonial power, and especially to missionaries.[40] Tradition could be used as a political instrument, against colonial powers, because it was "reified," as anthropologists say, that is, construed as external fact rather than a combination of concepts and preferences in the minds of many individuals.

Groups Are Not Like Agents

The belief that groups are like agents seems to be a straightforward consequence of the need to coordinate for collective action. To achieve specific goals, human groups engage in intense communication, which requires some description of common goals, as well as the means to achieve them. This in turn means that the motivations of members of the group are more easily communicated to others, and more likely to recruit more participants, if they are formulated as the wishes and beliefs of the group, that is, if the group itself is described as an agent.[41] If most of us agree that we should overthrow the tyrant, it is of course simpler to describe that fact as the common desire of the group, rather than as a complex concatenation of individual wishes.[42]

This belief in groups being like agents is easily entertained by human minds, because of the overwhelming importance of intuitive psychology in our mental life. Remember that a whole suite of specialized systems automatically picks up social information—other people's behaviors, gestures, utterances, but also their facial expressions, choice of words, and so forth—to construct, without any conscious effort, a representation of their beliefs, intentions, and emotional states, all things that cannot be observed and must be inferred. These systems, like other similar inference systems, are automatically activated when some information in the environment meets their input criteria, in this case information about human behavior.

So it is not too surprising that intuitive psychology systems are the main resources available to our minds when we try to describe and understand social groups, producing such thoughts as "the lineage wants this" and "the peasantry will need that." Obviously, the main cue that triggers thoughts about groups as persons is that they are actually composed of persons. So the move is not clearly perceptible between explaining the behavior of one or a few agents in terms of beliefs and intentions and extending this to a whole collection of agents, and then as institutions composed of agents, like corporations or lineages or social classes or kingdoms.

The belief that groups are like agents also generates all sorts of incoherent inferences or predictions, however. If we think that nations or other groups are persons, we may wrongly attribute to them memory, perception, reasoning, or other psychological processes. But social groups, organizations, and institutions do not really remember. Many veterans, for instance, have found to their surprise and to their chagrin that the nation that sent them in harm's way did not show much gratitude afterward. Also, the citizens of any modern large-scale society know of many examples of incoherent or inconsistent actions on the part of governments and bureaucracies. Even in small-scale communities, thinking of the lineage or the village as an agent leads to attributing to that group's mind beliefs and goals that may be inconsistent.

The problem goes deeper. When we reason about social processes, we often construe the groups as composed of generic agents, like "the poor" or "the employers." These generic descriptions seem to capture, in a simple way, features that are roughly true of most members of a social category. This is very misleading, however, as what happens to social categories or groups in many cases depends not on these generic properties but on the way preferences are distributed within a category.

Consider, for instance, the emergent processes that occur as a result of many individual decisions that had nothing to do with an intention to produce the overall pattern. A classic illustration is ethnic segregation in housing. Given the pattern of segregation in a particular city, it is tempting to draw straightforward conclusions about the attitudes of the individuals concerned. If there are ethnically diverse neighborhoods, we may think that they are inhabited by people generally tolerant of other ethnic groups. By contrast, a city where different groups are confined to different places, one would conclude that the inhabitants, perhaps those of one particular group, are strongly in favor of segregation. But the differences between individual attitudes in the two cities may actually be minimal, as the economist Thomas Schelling, and many other social scientists after him, pointed out.[43] A small degree of own-group preference can in fact generate highly segregated neighborhoods. So the emergent properties of the system provide little information about the preferences of individuals.

To complicate the matter even further, a description in terms of generic agents cannot accommodate the reciprocal influence of agents on each other, the crucial point being that the frequency of particular choices in a group affects the distribution of the preferences. This is the cascading phenomenon that I described in chapter 1, in relation to ethnic signaling. By choosing to demonstrate your ethnicity or your political choice, you change the frequency of such signals in other people's environments, and therefore change the apparent costs and benefits, for others, of adopting these behaviors.

Emergent effects of this kind occur in large-scale societies. But they are present also in smaller communities, as soon as interactions include more than a few dozen people. As an illustration, consider the dynamics of group conflict and conflict resolution in rural Morocco, as described by Ernest Gellner. Central to the interaction between the different tribes or factions are religious intermediaries, local "saints" known for their piety, wisdom, and close connection to God, described as *baraka,* roughly similar to "grace." These individuals are often credited with miracles, much to the dismay of orthodox, mostly urban 'ulema, who generally frown upon such superstition. The saints are also frequently involved in conflict resolution, as their baraka supposedly places them above and beyond disputes between families and tribes.

So having a few saints around is a necessary condition for peace and order. As saints come from specific families, supposedly related to the Prophet's lineage, it would seem that there should be an ample supply of such peacemakers. But only a few men stand out as real, functional saints. How do people select them? They are not, or so it seems, selected at all.

What happens is that some individuals from the right families behave in ways that make people guess they may be saints. In particular, they stand out as particularly generous, selfless, and hospitable. Interestingly, when individuals are identified as saints, they receive many gifts, as people want to cultivate them and perform charitable acts. These resources of course make it much easier for the saint to demonstrate generosity and hospitality, which further reinforces his identification as a real saint. The process is largely circular, but is not seen as such by the individuals concerned, as individuals cannot keep track of the pattern in other people's donations. So it seems to everyone that, through some mysterious process, a saint was elected even though no one deliberately participated in electing him. People attribute both the fact that there must be a limited supply of saints and the process whereby a selection occurs to divine intervention, clearly outside human agency.[44]

Descriptions of groups as generic agents ("the workers," "the landlords," and so on) also mislead us in a subtler way, by masking one central phenomenon of large-scale social dynamics, to do with the shape of a distribution of preferences. Here is why. A commonsense characteristic of agents, implicit in our intuitive psychology, is that they have specific preferences. For instance, we assume that each citizen has views about social welfare that fall somewhere between the two extremes of (a) judging all state help as unacceptable and (b) favoring state help for all cases of misfortune that may befall any residents in the country. To simplify their descriptions, political scientists often describe these various views as specific points on a line, between the extremes of 0 and 1, in this case total opposition and total support for welfare benefits, respectively. Any opinion on the matter is at some point on this abstract line.

This can help show how the description of social categories and groups as generic agents is severely misleading. That description would force us to consider whole collections of individuals (like "the workers") as being positioned somewhere on the line. But social processes follow very different patterns, depending on the way individuals are distributed along the different points of the line. If, for instance, almost everyone is at the .70 point of the curve (strong support for the policy), we can expect some consequences. But the consequences will be very different if half the population is at .99 (total support) and the other half at .41 (mild disapproval), even though this distribution yields the same average of .70. Another problem is that the preference curves for particular policies may peak at a certain value—that is, there are, for instance, more people at the .70 point of the spectrum of possible attitudes than at other points—but consequences differ a lot, depending on the shape of the curve. If it is very pointy, so to speak, almost everyone agrees with a certain preference, so the choice of policy is not a problem. If the curve is very flat, it means that the preferences of most individuals are far from the peak, from the most popular position.[45]

The description of a group in terms of generic agents does not just leave aside these crucial aspects of social dynamics—it makes it impossible to consider them. So the point here is not that generic descriptions of social categories or groups, typical of our folk sociology, are misleading because they oversimplify. That much is well understood, even by people who use such generic labels. No, the problem is deeper. By using generic labels for social categories and assuming that they describe common properties of agents, our folk sociology ignores the factor that initiates social dynamics, the fact that preferences are differently distributed among agents.

Power from Interaction

The limits of folk sociology are also visible in our spontaneous use of the "power as force" metaphor. Concepts of power describe the fact that one agent's preferences may in some circumstances prevail over another agent's preferences by directing the latter agent's behavior. So, for instance, the fact that the president wants people to wear hats of a certain shape results in them wearing those hats—as happened in Turkey as a result of the famous Hat Law of 1925. To mark the nation's evolution away from Ottoman Islam toward secular modernity, Mustafa Kemal's government decreed that men should abandon their traditional fez or turban and adopt Western headgear. There was strong opposition at first, mostly from rural areas and from religious traditionalists—but this was a severely authoritarian regime, so hundreds of people were arrested for violating the law, fifty-seven of them were executed, and the modern hat prevailed.[46]

This kind of process is what we usually describe in terms of forces. In this view, the combined "forces" of Kemal's government and a very efficient police were strong enough to "overcome resistance" from traditionalists who did not want to be "pushed around" but were eventually "crushed." The notion of power as a force seems self-evident to many human minds.

That is probably because our understanding of power recruits some cognitive resources that evolved for very different purposes. Among the many inference systems that compose a human mind, some are dedicated to describing the physical properties and behavior of solid objects—what is generally called an "intuitive physics" in the psychological literature.[47] Experimental studies show how infants, from an early age, spontaneously develop an understanding of objects in terms of solidity (objects collide, they do not go through one another), continuity (an object has continuous, not punctuated, existence in space and time), and support (unsupported objects fall).[48] Some aspects of intuitive physics are specific to humans. For instance, while chimpanzees' physical assumptions are grounded in perceptual generalizations, those of human infants seem based on hypotheses about underlying, invisible entities, such as forces or centers of mass.[49]

In the case of power relations, our conventional metaphors seem to recruit a specific subset of intuitive physical assumptions, to do with what the linguist Leonard Talmy called "force dynamics."[50] This denotes how our minds represent forces, motion, and interaction between solid objects in a way that is reflected in natural language. Force dynamics seems to rely on some kind of abstract mental picture of objects and their interaction, for instance an "agonist" that moves and triggers motion, resistance, or contrary motion in an "antagonist" object. This kind of dynamic is implicit in common descriptions of physical events, like "the brick broke the window" or "the ball kept rolling despite the stiff grass." As Talmy pointed out, these simplified schemas of possible interaction are implied not just in the way we express the motion of objects but also in many other domains of thought. In particular, we use force dynamics to express social influence, when we say that people are "pushed" by their peers to do something, that they did it "under pressure," and so forth.[51] Force dynamics is also involved in the common description of power relations in spatial terms, with one object weighing on another one, as when we say that some people have power "over" others who are "under" their control.[52] This is also clear in the quasi-universal notion that important, powerful individuals are on "top" of the political space and the powerless at the "bottom."[53]

But these physical terms are only metaphorical. The reason for which a Turkish man decided to wear a Western hat rather than a fez consisted not in abstractions like the "force" of authority but in his expected costs and benefits and his intuitions about other people's probable actions. After the law was enacted, he would guess there was some probability of being detected if he chose to wear the wrong kind of hat, some probability that others would denounce him to the police, some probability that the police officers would take him to jail, and so on. The combination of such probabilities and the potential costs is enough to sway even an ardent opponent toward obedience. The same, obviously, goes for each of the agents in question. The behavior of each policeman is constrained by his representation of what others would do to him if he broke ranks, for instance of the way his superiors would probably react. Each of his superiors could entertain similar intuitions about the costs and benefits associated with different courses of action—and so forth for all the individuals concerned, along a very long chain, or rather multiple chains, of contingencies that may include thousands or millions of individuals.

So the force dynamics that come to mind when we think about power relations, those notions of pushing and pulling, of force and resistance, are only very awkward ways of representing large-scale interactions that are vastly more complex, and indeed too complex for our conscious representations. In many situations, this metaphorical understanding of politics is sufficient. But it fails in crucial cases; in particular, it makes it very difficult to understand why some political systems endure, and why they suddenly collapse.

That is especially clear in the case of extraordinarily oppressive systems, like the European socialist regimes before 1990, that could survive for decades despite widespread popular rejection. The economist Timur

Kuran suggested that communication and coordination might explain this paradox. As many historians have pointed out, the communist regimes did not invariably resort to spectacular terror, as the Soviets had done in the 1920s and 1930s. Rather, in most cases the party operatives made sure that the population was involved in all sorts of demonstrations of adherence to the regime, participated in numerous parades and celebrations, applauded the leaders, proclaimed their communist faith, and inserted approved slogans in every activity, however seemingly remote from politics—Václav Havel described how Czech greengrocers would place a "Workers of the World, Unite!" sign in the middle of their carrots and onions.[54] People knew that they had to display commitment, and produce occasional statements of faith in socialism, to keep the police off their backs, as it were. So everyone participated in what everyone knew to be a farce. But the pervasive lying had one important consequence. It made it very difficult for anyone to gauge other people's level of commitment.[55] This process was pushed even further in places like East Germany, where the surveillance of the population was so extensive that virtually everyone could be suspected of having at some point collaborated with the infamous Stasi, the communist political police. Over forty years, the Stasi used threats and blackmail to recruit more than a million individuals as "unofficial collaborators" who occasionally provided information about minute details of their fellow citizens' behavior, such as who went with whom to what restaurant, what they ordered, and so forth.[56]

Authoritarian regimes do occasionally resort to exemplary, spectacular violence against dissidents to persuade the populace that resistance may be very dangerous. That is why the Pinochet regime in Chile used torture and the Videla junta in Argentina made thousands of opponents "disappear." But the need for such signals is much reduced if each individual can be persuaded that many other individuals around him might side with the regime.

So, in the opposition between a totalitarian regime and the populace, two separate dynamics take place. On the side of the regime, the apparatchiks and other minions maintain a reasonably efficient form of coordination. They have agreed objectives and mutual knowledge of each other's role in defending the regime. For example, party functionaries in communist regimes knew what their precise role was in the regime, they shared a precise understanding of what behaviors counted as dissidence and what measures were available to suppress it. On the people's side, by contrast, there are very few possibilities of exchanging signals of coordination. In the case of the communist regimes, although people massively despised the authorities, they did not have enough knowledge of each other's preferences, or knowledge of the appropriate time and place to express those preferences. To a certain extent, that is a general phenomenon in large societies, where most individuals are in what is called a situation of pluralistic ignorance. That is, they have much less information about other people's preferences than about their own. Many empirical studies support this point, showing that people tend to overestimate other individuals' commitment to the norm they follow.[57] But this is amplified in a repressive society, where communication between individuals is difficult and often dangerous. An individual may know that she takes part in the May Day parade as a way to avoid being hassled by the party minions, but all she knows of others is that they are participating, not the extent of their private discontent.

Coordination dynamics also help understand the sudden collapse of these regimes in 1989. The process was so sudden and so thorough that it stunned citizens, party bureaucrats, foreign observers, historians, and political scientists alike.[58] How could regimes that seemed so stable, backed by powerful and occasionally brutal police and armed forces, vanish in a matter of months or even weeks? Obviously, the processes that led to this dramatic result are much too complex to be described in a few lines—but some of the main features are instructive, as they show to what extent political power is, precisely, not at all similar to a physical force. As most historians point out, one crucial proximate factor was the Soviet policy of perestroika, which disrupted the coordination of apparatchiks. Because of the confusing signals sent by the central authorities, enforcers of the communist regimes did not have clear information anymore about each other's expected behaviors, especially in handling dissent and protest. The same dislocation of previous coordination occurred between the Soviet Union and its satellite regimes, as the Soviet authorities claimed to defend socialism in the

Eastern block yet also indicated that they might not participate in military suppression abroad, as they had in previous cases. Partly as a result of this loss of coordination among the authorities, people in various places felt more confident that they could demonstrate against the regime. The authorities in some cases tried to counter this with brutal repression, but in many other places the police response was weak or incoherent. An effect of these demonstrations was to provide people with information about support for the regime. As each individual could now detect that many others were in the opposition, this naturally led to each demonstration bringing together more people than the previous one, a process that was especially clear in East Germany. In other words, just as coordination was severely impaired on the side of the repressive authorities, it was becoming much easier on the people's side. These two movements led to bandwagon effects, as more and more of the regime apparatchiks guessed it was time to jump ship, while more and more ordinary citizens realized that the cost of protest was lower than ever before.[59]

Our intuitive folk psychology does not have the tools to represent those dynamics, to explain in this case how communist regimes could both survive for decades when most people were against them, and collapse within a few months. If we want to describe political power as a physical force, we have to assume that communist leaders for a long time just had that force, that mana as it were, and that it suddenly left them afterward. But that is clearly inadequate. The problem of the force-as-power metaphor is that, like the groups-as-agents metaphor, it makes it difficult to consider the domain where political power is actually created, in the extraordinarily complex aggregation of interactions between individuals.

Norms in the Heads

The complexity of interaction is also the reason we find it so natural and compelling to think of social norms as things, as external to the many minds that hold some representation of the norm. This idea that norms are objective seems entrenched in our ways of thinking about social relations, and this is manifest even in childhood. A long time ago, Jean Piaget had noticed that children tended to be "moral realists," to consider moral norms, but also some conventions, as a matter of objective fact. From this perspective, it is wrong to hit other people, even if no one has said or thought that it is wrong.[60] Studies by Elliot Turiel and others then showed that children are in fact more sophisticated moral thinkers, as they draw a distinction between moral rules, on the one hand (for example, one should not hit others without provocation), which in their view are valid whether or not they were made explicit, and social conventions, on the other (for example, only women wear skirts), which must be explicitly prescribed.[61] But children also understand the normative force of conventions, by which people feel they ought to follow them. Children (and adults too) are aware that norms change from place to place and from time to time—but it seems that they also expect norms to be social realities, independent of people's ideas.[62] Indeed, more recent research, beyond the domain of morality, shows that children, from an early age, have definite expectations about the normative character of social conventions. For instance, preschoolers consider the rules of a newly invented game to have normative force, and they protest loudly if others try to play along different rules. They consider such changes impermissible, even when the interaction in the new game was presented in descriptive ("one does this, then that") rather than normative ("one must do this, then that") language. Even more striking, young children spontaneously switch from descriptive to normative language when explaining the activity to newcomers. This is despite acknowledging that norms and rules are, to some extent, arbitrary and that they may differ between communities.[63] These experimental studies suggest that, from an early age, we can learn about the specific norms of our community, on the basis of prior intuitive expectations about what norms are and how they apply to behavior, and these expectations seem congruent with the idea of norms as both compelling and objective.

Now, of course, social scientists would say that this is largely an illusion. Social norms constitute a type of convention that supports coordination.[64] People's representation of a regular behavior (for example, people shake hands when being introduced) becomes a norm if they also represent that (a) people should shake hands when introduced and (b) others expect each other to abide by that rule. Once a certain number of individuals in a social environment have these two representations, there is potential coordination.[65] Obviously, this is a very abstract summary of people's potential representations, which in actual cases are filled with far more specific content about what one should do, how others might react, and so on.[66]

The discrepancy between the way we spontaneously construe norms, as objective rules outside people's heads, and the way our interactions produce coordination is the reason we often are baffled, in our folk-sociological reasoning, by the power of norms and the fact that they can change. In the same way as for political power, our folk sociology makes sudden transformations of the social world appear mysterious. A good example is that of foot binding among the Chinese upper classes, a practice that for centuries was considered not just normal but imperative and embedded in traditional Chinese values—and disappeared without trace in a few years. This is an example of "moral revolutions" that occur without dramatic political or economic upheavals, as a result of subtle changes in people's interactions.[67]

For centuries, it would have been unthinkable for most Chinese aristocrats, mandarins, and other upper-class people not to bind their daughters' feet, thereby making them unable to walk, let alone work. People were well aware of the pain endured by young girls and of the risks to their health. But the practice was entrenched. It was congruent with other representations of proper gender roles among the upper classes. It was thought to be one of the many interlocking norms that constituted Chinese tradition.[68] It even affected people's criteria for sexual attraction, as some authors describe the erotic passion aroused in male characters by the sight of stunted feet.[69] Yet the practice disappeared very quickly after the inception of the Republic of China in 1912.

How can we make sense of this contrast between the apparent solidity and the sudden fragility of a norm? Describing norms as something "out there" as a social fact that is independent of people's representations certainly does not help here. Why would the social fact suddenly disappear? A more promising perspective is to consider individual representations and motivations, and the way they are combined and produce interaction in a large-scale society. A norm is mentally represented as combining information to the effect that, for instance, "feet should be bound" and that "(somehow) everyone (in the relevant group) agrees with this rule." From these pieces of information, people can infer further representations, for example, "everyone (in the group) will disapprove of us if we stop doing it." So everyone acts as if the behavior in question was actually a norm, which makes it a norm.

But the recursive process of coordination itself is entirely opaque to people who follow a norm. The reason foot binding was, for centuries, an imperative norm among the Chinese upper classes was the particular way in which these representations about (a) the rule and (b) its prevalence among other people were distributed among thousands or millions of different minds. In one's social environment, it seems that everyone is following the norm. Note that people's representation of "everyone" is usually very vague (one does not need to figure out even in rough terms how many people hold the norm) and usually silent about other individuals' commitment to the norm, as I mentioned above. However vague, this "everyone" representation is very efficient, as it allows people to predict what would happen if they changed their behavior. If they stopped following the norm, the neighbors would disapprove and perhaps punish them.

The prediction is usually accurate—when a norm is accepted, people do disapprove of violators. But it is accurate for the wrong reason. Neighbors will disapprove, and in many cases express disapproval, not (just) because they believe in the necessity or usefulness of the norm but also because the costs of approving a violator may be high, and the benefit from being seen as an upholder of norms may be significant. Now these costs and benefits for the neighbors are determined by the fact that they themselves have neighbors . . . whose costs and benefits for approval or disapproval or norm following and norm violation are created by their own neighbors, and so forth.

It is almost impossible for a human mind to represent such chains of recursive causation—note how even the simplified description above required ponderous prose and numerous repetitions. It requires vast computation to track in explicit terms the multiple contingencies that result in apparently organized behavior, the fact that a's behavior depends on his representation of what $b, c, \ldots n$ will do or are doing, behaviors that themselves depend on representation of each other's behavior, and so forth. But these multiple contingencies are the real process here.

Only interaction patterns can explain how the norm can in some cases disappear very easily, even though it seemed imperative a few years or decades earlier. In the case of foot binding, all it took was for a number of people to engage in fairly simple forms of collective action. During the nineteenth century, various Chinese authors as well as foreign missionaries had popularized the idea that foot binding was after all optional, perhaps an arbitrary choice that was not as imperative as most people thought. But these efforts had little result. Finally, in the first decades of the twentieth century, modernist families started the movement against foot binding by establishing "Healthy Foot" associations founded on a double pact. Their members would not bind their daughters' feet, and they would all forbid their sons to marry a foot-bound woman. The movement snowballed, and within a few years the practice had been all but abandoned in most of China.[70]

In terms of game theory, of the abstract formulation of strategic moves and countermoves, this kind of collective pact constitutes an optimal strategy. First, many individuals and families get together to vow they will abandon the traditional norm. Once there is a sufficient number of reformers, they cannot be victimized or ostracized as isolated rebels would be. Second, that very fact lowers the cost of joining, for outsiders who agreed with the need for change, but worried about social disapproval. So membership in the new covenant increases. Third, the families make that commitment public, which makes it difficult for any members of the association to defect, to return to the old practice, on pain of losing their reputation for trustworthiness. Fourth, the promise to take the members' children as brides and grooms (in some cases to marry only members of the association) further reduces the cost of involvement. That again makes participation more tempting for outsiders. Given all these conditions, one should expect membership to grow, all the more so as each increase in membership makes joining less costly.

Folk Sociology as a Coordination Tool

The assumptions of folk sociology (that groups are agents, that power is a force, that social facts are things) are very similar to what linguists call "conventional metaphors." Examples include such common tropes as "time is a resource" or "debate is warfare." People say that they "wasted" or "invested" time, and accounts of intellectual disputes abound in terms like "winning the argument," "defending a position," and so on. This is clearly metaphorical, as people do not literally believe that they can "crush" an intellectual adversary as you crush garlic. Nor do they, one hopes, confuse "attacking" an opinion with beating up an actual person. But the metaphor organizes their thoughts about debate. As George Lakoff and Mark Johnson pointed out, such metaphorical understandings are pervasive in human cognition.[71]

Why do we resort to such conventional metaphors when we represent our social world? As I described above, one crucial factor is that a more realistic description, in terms of interactions between individuals, is simply beyond our capacities. The aggregation of myriad individual behaviors, many of which are prompted by other agents' behaviors, constitute complex systems, beyond what human minds can represent in consciously accessible form.[72] In other words, we could say that we are condemned to use folk sociology, with its misleading assumptions, because of the mysteries of coordination—the mysteries of apparent order created by the aggregation of myriad interactions that we cannot follow.

But that is not the only reason. Indeed, the ways in which our spontaneous folk-sociological reflections make sense of the social environment also constitute very efficient coordination tools. Coordination is in many cases made much easier by the fact that most human minds entertain very similar representations of the social world, what I called their folk sociology, even though those representations are not entirely accurate. For instance, the assumption that groups are like agents, although it is misleading in many cases, also serves as a coordination device. People can entertain a summary description of their own group and others, in a way that (roughly) accounts for different agents' behavior, and directs their own behavior.[73] In this sense, the belief constitutes a self-fulfilling prophecy. To the extent that most people believe in the description of the group as having intentions and beliefs, the behaviors of individual members tend to (roughly) confirm that assumption. The same is true, obviously, of the notion that norms are external to people's minds. To the extent, for instance, that most Fang people assume that there is a proper way to perform the ngam ritual, and that these rules are somehow external to people's minds, it is easier to agree on past performance as the guide for carrying out the operation.[74] In the same way, people who understand power differentials as similar to physical dynamics end up behaving in ways that (roughly) correspond to what a metaphorical description in terms of forces and motion would predict.

Such effects are well known to social scientists. They are also the reason why, in their description of particular groups, nations, or institutions, social scientists often tend to adopt the language of folk sociology, and describe groups as quasi-agents, or social norms as a form of "culture" that is independent of people's minds. These approximations roughly fit the evidence and provide a convenient shorthand for coordination processes that would be exceedingly difficult or cumbersome to describe otherwise. But folk sociology is not good social science.

Lessons for Modern Politics?

Understanding the assumptions and the limits of our spontaneous folk sociology is not just of academic interest. Figuring out how we managed to scale up human groups, from small bands of foragers to large industrial societies, is crucial to making sense of modern politics. One lesson from the study of folk sociology is that our political psychology, like the rest of our cognitive functioning, consists for the most part of implicit processes to which we have no conscious access. That is, if we want to understand what mechanisms lead people to find particular programs or policies compelling, or particular leaders worthy of support, we must not limit ourselves to explicitly held, conscious opinions and reasoning.[75] Indeed, many political scientists have recently turned to the systematic study of the implicit cognitive processes involved in political choices.

People in modern large-scale societies are presented with platforms, that is, set menus of policies. One must accept the whole menu or switch to a different one. In many modern democracies, the commonly available menus are called conservative and liberal. Often, citizens find one particular menu more appealing than the other, even though it is not clear why distinct policies are assembled in those particular packages. Why is a relaxed morality generally combined with a preference for high progressive taxation? Why would a pro-business attitude entail attachment to conservative morality?

What makes each political package hold together, in terms of explicit ideologies, seems to be very abstract values, or the ranking of different values, such as liberty above equality. In broader terms, the economist Thomas Sowell proposed that the modern Western opposition of conservative versus liberal mind-sets is associated with two fundamentally different visions of society and human nature. One such perspective is a "constrained" vision, common to Adam Smith, classical liberalism, Burkean conservatism, and many modern-day libertarians. People who adopt this vision accept that human nature is imperfect and will ever remain so. Policies should be piecemeal, pragmatic palliatives to specific problems rather than grand schemes for a better society. By contrast, many progressives spontaneously converge on an "unconstrained" vision, according to which humankind is perfectible, and human misery is the outcome of imperfect social conditions. So policies should be conceived as steps toward the construction of that better society, in which the essentially beneficial features of human nature will unfold.[76]

But abstract values are not the only factor in the opposition between the conservative and progressive packages. The opposition also corresponds to a difference in moral outlooks. The psychologist Jonathan Haidt documented the different ways people apply moral intuitions and feelings to various domains of behavior. Many people, especially intellectuals, routinely understand morality as principles and intuitions to do with caring for others, not harming them, and with being fair. It is clear to them that our moral psychology is what motivates people to seek fair transactions and explains our disgust at unprovoked violence. But people have moral feelings and principles about many things besides harm and fairness. For one thing, people in most societies value loyalty to one's own group, and treat defection and desertion as despicable. Also, in many places people will think that eating a particular food on a particular day is immoral. Others will find it morally reprehensible to ignore or resist tradition and authority. This suggests that our moral intuitions can apply to other domains of behavior beyond harm and fairness, in particular, to those values Haidt characterizes as loyalty, authority, and purity.[77] Now one remarkable finding of Haidt's studies is that modern liberals and conservatives seem to differ in the scope of their moral intuitions. In liberals' minds, only harm and fairness are clearly moral. But offenses against tradition or authority are not considered in moral terms at all. Conservatives by contrast seem to have a broader moral sense, which is applied not just to harm and fairness but also to possible violations of loyalty (for example, people not being patriotic), of authority (failing to respect older generations), and sanctity (burning the flag). This would explain why debates between modern liberals and conservatives are generally intractable, as each side does not quite perceive the moral motivations behind the other's positions.[78]

Differences between these visions of society and politics seem to go even deeper than moral understandings. In a variety of experimental tests, apparently unrelated to political persuasion, conservatives and liberals behave very differently. Conservatives react more to sudden or surprising stimuli, and they seem more sensitive to potential threats. Their attention is more strongly mobilized by negative words. Liberals, more than conservatives, orient their attention where another individual seems to be looking—conservatives are less easily influenced. These differences are not just an American oddity, either, as experimental studies yield similar results in other modern countries.[79]

Can we make sense of these experimental results? Perhaps not in the present state of political psychology, especially not if we stick to the opposition between modern conservatives and progressives. This opposition is certainly too ethnocentric or modernity-centric, as it were. Modern deliberative democracy is a very recent idea, and an even more recent practice in most of the world. Also, it is only in some countries, at some times, that political debates take the form of an opposition between conservative and progressive agendas. In many places, ethnic rivalry is a more salient cleavage line than attitudes to redistribution and state intervention in the economy. So linking genetics and personality to modern political choices is bound to be awkward, as the terms being compared occur on such different timescales. Whatever genetic differences there are between individuals, and the personality differences that they contribute to create, have been around for a long time, for tens of thousands of years and probably much longer. They appeared, and were selected, in human groups that had a very specific, small-scale political organization. So perhaps we should look for explanations in more archaic assumptions from folk sociology, and in the way they are applied to modern contexts.

The State as Moral Agent

A major point of contention between rival political visions is the extent to which all manner of social functions should be provided, organized, or regulated by the state. For instance, at the onset of the American republic, the founding fathers stipulated that the federal government should provide a postal service, but not unemployment insurance. Many Americans these days would consider the opposite choice more sensible. True, most people agree on the minimal functions of the state, such as national defense, the protection of citizens from theft and assault, and the institution of fair, law-bound courts to ensure the execution of contracts. But modern states do vastly more than that, and the debate revolves around the question, How much of that additional activity is legitimate or desirable, and how much is unnecessary or downright detrimental to general welfare? Obviously, this is not the place to adjudicate on the substance of that question. Rather, we should consider how our evolved capacities and preferences give a particular shape to that debate, what tacit assumptions are taken on board when people discuss the proper functions of the state.

One important psychological factor is that many people represent the state as a form of insurance against the vicissitudes of market processes. To guarantee against the worst consequences of failure, it would seem reasonable to invest in some insurance scheme, and the state may be seen to provide just that. That is why in many countries, notably in Western Europe, many people consider residual social welfare programs—safety nets for those whom the market failed—to be indispensable in any modern society, just like national security and the protection of life and property.

This, however, still does not explain why the state itself would be seen as an efficient buffer against uncertainty. After all, self-organized communities can also provide insurance against all manner of uncertainty, as in the case of the voluntary associations that emerged in newly industrial countries in the eighteenth and nineteenth centuries. One consequence of the industrial revolution was that many people, having left their villages, had to provide for each other the kind of solidarity that was previously provided by kith and kin. They formed associations to deliver services that had been the preserve of the upper classes, like access to medicine and education, and to provide invalidity or unemployment insurance. In England alone, there were hundreds of such "friendly societies" by the beginning of the nineteenth century.[80] These were mostly self-organized associations of a few hundred or a few thousand individuals, with offices held in rotation by voluntary members. Many friendly societies employed a physician on a retainer. Some of them, like the famous Oddfellows (the society for "odd" people who did not belong to any particular trade or occupation, and therefore did not have their own friendly society), organized lectures and published magazines to educate their members.[81] State authorities looked askance at these spontaneous organizations, seeing them as potential fomenters of revolutionary enthusiasm. Indeed, an important motivation in the development of state-provided insurance and medical services was a push against these bottom-up associations.[82] So the modern preference for state provision may reflect a political situation in which state bureaucracies managed to ensure a monopoly.

The fact that the state can easily be seen as the most likely provider is congruent with our folk-sociological assumptions. When people represent the role of the state in public affairs, they generally construe the state as an agent, in keeping with the most common assumption in folk sociology. In the same way groups or social categories are often mistaken for quasi-persons, the state is described as an entity that has knowledge and intentions. This description is of course misleading, as there is no central repository of knowledge for the state, no central intentions and memories.

Rather, there are vast numbers of interactions between individuals, the complexity of which greatly exceeds our cognitive capacities (and indeed exceeds the power of most scientific models). So the human mind is placed in a situation in which unfathomable processes (the complex reality of decision making in a modern state) result in people actually doing things, such as bureaucrats distributing benefits, tax officers demanding payment, police officers arresting people, and so forth. The easiest way for a human mind to explain what is done by many human actors in a complex set of interactions is to summarize all this complexity as the doing of one generic or single mind, the mind of the state.

This compelling description of the state as an agent is bound to be approximate and often wrong, as modern bureaucracies comprise thousands of agents, human agents, who as it happens have their own goals and intentions and knowledge, as well as their incentives, which may or may not line up with the goals of a particular policy. Indeed, a whole field of political economy, public choice theory, is concerned with this question of the workings of the state, once we understand that it comprises many different agencies, each of which includes many agents, all of whom are faced with particular incentives for behavior—and these incentives may or may not align with the mandate given to political parties and government.[83]

The spontaneous and compelling belief that the state is like an agent may also explain another feature of modern politics, the fact that policy choice is often driven by moral intuitions, particularly about the motivations behind particular policy proposals. Politicians generally present policies as ways to remedy particular problems, for example, to provide better schools, more extensive health care, higher wages, more employment, and so on. Debates about policies also revolve around these declared intentions—people, for instance, argue whether we should make schooling or health care a higher priority than wages or welfare benefits. Economists often lament this focus on intentions at the expense of results, which they see as particularly damaging in modern economies, where policy choices have important unintended consequences and often perverse effects. So why would people focus on intentions and moralized descriptions of political programs?[84]

This may be a result of the belief that the state is an agent, and of our intuitions about economic exchange. As I described in a previous chapter, it makes sense for humans, in their evaluation of potential transactions, to gather information about intentions as well as the current offer. That is because in premarket conditions exchange was most advantageous when it consisted of potentially reiterated transfers for long-term mutual benefit, rather than one-shot impersonal interactions. In such situations, it was often more important to identify and select a well-meaning partner than to choose the best deal. A potential partner's intentions, inasmuch as one could detect or infer them, often provided a reliable guide to future benefits. Which is why we now spontaneously find such information of great interest, even in market transactions where it may not be as relevant as it was in ancestral conditions.

So many citizens of mass societies may focus on the intentions behind policies, and on a moral evolution of these intentions, because they treat the state like an agent, that is, an entity with goals and beliefs. Given this tacit assumption, it is natural to construe the services provided by the state, and the duties it imposes, in the same light as the terms of exchange with a human partner. Political ideologies in mass societies often reflect this implicit representation. Social democrats construe the state as a mostly benevolent distributor of deserved benefits, with taxes as the counterpart in a fair exchange. Some conservatives and most libertarians see the state as an exploitative partner, whose enormous resources and monopoly of violence clearly predict unfair exchange and disastrous outcomes. Both visions are grounded in our evolved folk sociology, which partly blinds us to the actual workings of state institutions.

Deliberation among Evolved Minds

In the tiny republic of Te, politics is mostly transparent. When people are discussing the proper allocation of grazing commons or when they decide how to organize the repair of dams and construction of canals, they are aware of the various intentions of different people, of their interests, and of the consequences of each decision for each member of the community.[85]

That transparency is out of reach in mass societies. The ideal of deliberation is that some of that transparency can be regained once we establish proper institutions for communication and decision making. But is this compatible with human political psychology?

In a sense, the emphasis on deliberation is consistent with our capacities. Deliberation is made possible by our evolved reasoning capacities, and this explains why, as historians and political scientists have long observed, free and open deliberation generally leads to choices superior to those of autocratic and technocratic systems. As I mentioned before, reasoning is activated mostly for the purpose of argumentation. Bluntly, it seems that we do not reason in order to get a more accurate picture of the world. We reason primarily to bring others around to our beliefs and preferences. That is what makes so much reasoning self-serving, but also, surprisingly, makes discussion the best approach to decision making, as we are much less vulnerable to other people's bad reasons than to our own.[86] So our evolved cognitive dispositions for "epistemic vigilance," the detection of deception and error, are fundamental to deliberative democracy. However, deliberation, to be efficient, requires not only that we have reasons for or against policies but also that we are aware of the causes of our preferences, and that we have a roughly accurate picture of the way social processes work. But we rarely meet these two strict requirements.

What is to be done? Obviously, the study of the political mind does not by itself translate into policy recommendations. But it could help us bypass our entrenched notions of how societies work, our folk sociology. It could also lead to a different vision of the political debate, one where we can use what we know of evolved human capacities and dispositions— concerning, among many other domains, the motivation to form coalitions, the disposition to form families, the propensity to strange beliefs, the urge to invest in kin and offspring, and the capacity for extensive cooperation.

Conclusion

Cognition and Communication Create Traditions

HUMANS STAND APART FROM OTHER species in the amount and diversity of information they acquire by paying attention to other humans' behavior, to what others do, and, crucially, to what they say. It is difficult for us to realize how much information is socially transmitted, because the amount is staggering and the process is largely transparent. There is an ocean, a mountain, a continent—such metaphors are all apt and all mislead-ing—at any rate, an extraordinarily large amount of information that is being transferred between people, at any point, however small the community. Information is our environment, our niche, and as we are complex animals we constantly transform that niche, sometimes in ways that make it possible to acquire even more information from our surroundings.[1]

This metaphorical ocean of information is where we find what people call human culture, or the different human cultures. Those terms, it must be said, are very vague and hugely misleading, and their use often led to com-plete confusion in classical anthropological theories. That is because the terms almost inevitably carry implications about information and human psychology that happen to be quite clearly on the wrong track—a problem I discuss below. Fortunately, we do not need to start with a definition of these words in order to ask meaningful questions about the transmission of information in human societies—just as you need no clear definition of matter to ask meaningful questions in physics, or a definition of life for biological questions. So here are some questions about information and its transmission.

245

In all human communities, people seem to "share" some mental representations. I use quotation marks around "share" because people of course do not share them in the same sense as we can share a meal. What we mean is that the representations in different minds seem to have some similar features. (That ambiguity about sharing is the first of many confusions created by terms like "culture.") To reprise an example from previous chapters, the notion that children are members of their maternal uncle's clan but not of their father's is found in a roughly similar form among people who happen to live in a matrilineal society like the Ashanti. Or the idea that shamans have a specific internal quality that makes them different from ordinary people, as commonly found among Fang people in Gabon.

This raises the question, Why are there such similarities in people's representations? This is where we run into another confusion created by the notions of culture or cultures. Because we have a name for something, we may be tempted to think that it actually is a thing, a coherent set of realities. That would be bad enough in this case. But, even worse, we may be tempted to think that the term by itself is an explanation, that the Ashanti assign clan membership though the maternal line because it is in their culture, as people sometimes say. But that obviously cannot be an explanation. Saying that the norm is part of Ashanti culture is tantamount to noticing that the representation of the norm is similar in the minds of many Ashanti people. The would-be explanation is circular.

So the question, Why do people who communicate have similar representations? points to a real problem. Among the multitudes of mental representations that a human mind entertains in the course of ordinary behavior, only a minuscule proportion are similar to other individuals' representations. We constantly build and update representations of our physical environment that are of course unique in some respects, as each individual has a unique perspective on the surrounding objects. We have representations of the social world around us that are also unique, since we are each the center of many networks of social relations, and nobody else occupies that particular position. Just as obvious, communication does not automatically create similarity of representations, nor is it intended to do so. If you make a request, you do not intend listeners to entertain that very same request. You want them to entertain a motivation to satisfy your request. Even when we use declarative statements we do not create mental representations similar to our own. If you state that roasted pangolin is delicious, that does not make your listeners think that it is delicious, it only makes them think that you seem to think so. All that may seem obvious, but sadly it is necessary to mention these familiar properties of communication, as they are often difficult to keep in mind, so dense is the fog created by notions like "culture."

Another question is, Why are there recurrent features in these representations in many different human communities? Consider the notion that shamans or other healers have some special substance or quality or additional organ. This is found, in a rich tapestry of different forms, in various places in Asia, the Americas, and sub-Saharan Africa. Why do people in such different places home in on this notion of internal essence? Or consider the idea that you belong to your mother's lineage and her brother's but not to your father's. That is common to Trobriand islanders in the Pacific and the Senufo in West Africa, as well as the Hopi of North America and the Nayar of India, and many others. Obviously, the same could be said of almost any feature of local norms and ideas reported by anthropologists. For instance, in many places in the world you will encounter the idea that social groups are different for natural, essential reasons—that, for example, ethnic categories correspond to different species of humans, or that individuals from different castes do not have the same physiological nature. Or people in many places have some notion that deceased individuals still exist as persons, despite physiological death, and can interact with the living. And people in many places consider that malevolence and magic explain misfortune better than random contingencies.

None of these is, at least at the surface, a universal human representation. You can have healers without a special organ—that is the case for herbalists in many African societies, and of course of modern physicians. You can have a society without lineages, like bands of nomadic foragers or modern mass societies. You can have social categories without essentialism. You can have notions of death without a surviving spirit, and of misfortune without agency. But even if they are not universal, these representations are remarkably frequent in human groups. Why is that the case? What explains cultural recurrence?

So, dispensing for the moment with confusing notions of culture, we have two questions for a natural science of societies, namely, How do people converge on similar representations through communication? and Why are some themes so common in such diverse, unrelated societies? At the risk of ruining the surprise, I should reveal that these are in fact one and the same question, which we can address in a rigorous manner by considering the way human minds infer new representations from communication.

Traditions

Let me start with similarity within a community. Some small parts of the immense domain of mental representations entertained by people in a group seem to be roughly similar in the minds of different people, as a result of past communication episodes. A convenient term for these islands of similarity in an ocean of unique content is "traditions." These are simply sets of mental representations and associated behaviors that have some stability in a particular social group.[2] Contrary to what the term may suggest, traditions in this sense may or may not last. Some traditions persist for centuries while involving very few individuals, like the art of choosing and mixing registers of different timbers on the organ or the art of making Kabuki theater costumes in Japan. Short traditions include all those fashions that quickly spread and vanish even faster, like so many flashes in the cultural pan. As these examples suggest, the duration of a tradition and the number of people involved are orthogonal dimensions. Traditions may be established between a few individuals, or they could reach millions. The writer Natalia Ginzburg provided a good illustration of a very small-scale tradition in her book *The Family Lexicon,* listing a number of odd words or usages used by her parents and siblings in prewar Italy. The father in particular had coined (or acquired from distant relatives) a series of partly invented terms as well as mixtures of German and Italian.[3] Many families have such small-scale traditions, even though these often reduce to a handful of special words or idioms. At the other end of the spectrum, some traditions are held by thousands or millions of individuals. Long before modern communication techniques made extensive diffusion possible, millions of people had acquired and transmitted folktales like *Cinderella* or the legends of Krishna and Ganesh, melodies like "Greensleeves" and hairstyles like the Manchu queue, not to mention largely widespread religious beliefs. So, when we say that these representations are roughly stable in a social group, we should understand this in the most flexible way. Any collection of people who are connected by some communication episodes can be said to have created a tradition, if those events result in roughly shared representations. This is a very broad and extensive understanding of traditions, but it should be sufficient to point to many problems with our common assumptions about information and transmission.

In particular, we may miss some crucial aspects of cultural transmission because our folk sociology, and the many social science theories that reflect its assumptions, expect widespread representations to persist. In that view, only change requires a special explanation. Social scientists for a long time assumed that there was nothing special to explain in the fact that many Venetian and Xhosa customs or ideas were very similar to what the Venetians and the Xhosa of the previous generation had been doing or thinking. In this perspective, one expects the stories of *Cinderella* and *The Monkey King* to be transmitted from generation to generation—the fact of stability does not require any special cause, as if there was some cultural equivalent to the law of inertia for physical objects.

But it is stability that is mysterious. Most of our utterances are not recalled by our interlocutors, most of what they recall is drastically edited, and the part that they may transmit to others depends on their motivations as well as myriad other factors—so that the ocean of information transmitted by human beings is a place of high entropy. That question had been raised by some scholars, notably Gabriel Tarde at the beginning of the twentieth century. Tarde had tried to explain how large-scale cultural effects, for example, the diffusion of a fashion or a political ideology, would result from the aggregation of many interactions between individuals, and in particular in the proportion of imitation and innovation in the way each individual's behavior was shaped by that of others.[4] So the difficult problem for social scientists would be to explain how individual encounters led to large-scale social effects. Unfortunately, these ideas had very little effect on the social sciences at the time, so that few people realized that, in the face of entropic communication, the improbable stability or spread of some representations is the true mystery, the phenomenon to explain.

Transmission as Selection

The transmission of cultural information was not properly studied in classical social sciences, at least not before the 1970s. Things started to change when biologists and anthropologists began to consider cultural transmission as a population phenomenon. The inspiration came from evolutionary biology, which had shown how the evolution of species, a large-scale phenomenon, could be explained by the aggregation of small changes in the replication of genes in individuals. Since culture is just the name of large-scale properties of information held by many individuals, and the only process whereby information is processed occurs inside individual brains, it seemed possible to propose for culture an analogue of biological evolution.

The most important development in this respect was the publication of Rob Boyd and Peter Richerson's *Culture and the Evolutionary Process,* which for the first time proposed clear theoretical tools for describing and explaining cultural processes.[5] Although there were antecedents in this view of culture as composed of different units with a different likelihood of transmission, these were not as systematic, nor were they as usable as the model provided by Boyd and Richerson.[6]

The starting point of the model was that cultural material comes in different packets of information, called memes, transmitted from individual to individual. The notion of memes had originally been proposed by Richard Dawkins, and it then formed the starting point of many attempts to describe cultural material.[7] In this selectionist perspective, trends in cultural evolution, for instance, the persistence of a particular tradition or its downfall, the fact that some ideas can diffuse to large communities or on the contrary remain confined to a few individuals, all stem from the relative selective success of different memes. This was a transposition to cultural material of the successful model of genetic evolution, based on mutation and selective retention.[8]

The theory described biases likely to affect transmission, that is, to push the evolution of cultural materials in a particular direction. The frequency bias, in analogy with frequency-dependent selection in genetics, specified that, all else being equal, more frequent memes would be more likely to survive transmission than rare ones. In other words, individuals who could perceive the difference in frequency between two memes would be likely to pass along the more frequent ones—in what could be called conformist transmission. This conformist attitude would make sense in a species in which a vast amount of crucial information is acquired from others, rather than from direct experience. In effect, trusting the more frequent meme, for example, cooking your stew the same way most people do, amounts to taking advantage of previous generations' trial-and-error-driven progress toward an optimal solution. Another factor in transmission would be a prestige bias, whereby we tend to adopt the memes (ideas, practices, ways of doing and communicating) of successful individuals. Even in groups with simple technology there are differences in the efficiency of particular gestures or methods, as in hunting or fishing or toolmaking. Finally, the formal model included what were called content biases, resulting from the fact that the human mind is predisposed to acquire or communicate some representations more easily than others, which would, for instance, explain why some tunes (like "Pop Goes the Weasel") are acquired easily while others (like Alban Berg's Lyric Suite) are not, in the same way as the plot of *Cinderella* is easier to narrate than that of James Joyce's *Ulysses*.[9]

This dual-inheritance theory, using mathematical models from population genetics to describe cultural transmission, allowed anthropologists and archaeologists to formulate hypotheses about transmission with much more precision than traditional social science models had provided. There were of course some problems in the application of the models. For instance, the prestige and frequency effects predicted by the model only occur some of the time. People may imitate the upper classes in etiquette, but they just as often adopt the accent or vocabulary of the *vulgum pecus*.

Simple frequency biases would predict bandwagon effects that sometimes happen, and just as often fail to occur.

A more difficult problem was the process of transmission. The model assumed that imitation was the main path of information transmission—a process whereby a mental representation would be created by copying some observed behavior.[10] This assumption was in fact shared by most models of culture based on a notion of memes, from Dawkins on to the later developments of what some people called "memetics."[11] At first sight, imitation seems a simple enough process. An individual hears "Pop Goes the Weasel" and forms a mental representation of the series of pitches and rhythms that constitute the tune as it was sung, whistled, or played. This representation then produces behaviors like whistling or singing a series of notes that (roughly) replicate the original performance. But even this simple example suggests that imitation is far more complicated than it seems.

First, obviously, imitation does not explain why some tunes are easier to replicate than others, why "Pop Goes the Weasel" is an easier meme to transmit than the first violin line in the Lyric Suite, which is not really more complex. That is why Boyd and Richerson had found it necessary to include "content biases" in their model. But these biases were left as an empty placeholder. The theory specified that they must exist, but it did not say anything about what they consisted in (as that was not the point of the dual-inherence model).

Second, even in the transmission of a humble and simple tune we find another feature of cultural transmission that is often ignored but happens to be of major importance, namely, that people's mental representation some-times turns out to be better, so to speak, than the material supposedly imi-tated. People who hear "Pop Goes" played on an out-of-tune piano or sung by a tone-deaf amateur can still mentally represent it as the correctly pitched series of notes, and if they have the relevant skills, they can produce a more melodically correct version of the tune. So the process of transmission does not reduce to imitation. It ignores some actual properties of a model and is guided by something other than what was observed.

These two problems with imitation are related. The reason some elements of behavior become traditions and others do not is that creating a tradition does not really consist in imitation but includes the constant re-construction and correction of input. This is all to do with those content biases predicted by the dual-inheritance model—among which we should count most of the dispositions and preferences documented in the previous six chapters. People are more likely to hold concepts of superhuman agents than of other possible supernatural notions. Rumors spread quickly when they describe potential threats. Descriptions of a complex economy are fil-tered through the templates provided by our intuitive sense of fairness. Ste-reotypes associated with social groups spread more easily if the group is construed as essential—it is in the nature of the members, so to speak, to do what they do—than if it is explained as mere accident. To complete this model of transmission, we need to include all these elements. In sort, we need much, much more psychology.

An Example: Social Essentialism

Ethnic groups and many other social identities are construed in terms of natural differences. That is, people think there is some real, internal differ-ence between groups such that members of group A are not and could not possibly be members of group B. One is a member of a particular group from birth, on the basis of inheritance.[12] For instance, the Mongol nomads interviewed by the anthropologist Francisco Gil-White were quite clear that Mongols and Kazakhs are different kinds of people, that it is not just an accident that they belong to different groups. For the Mongols, what makes each Kazakh person a Kazakh is not just that he speaks the language, or appreciates Kazakh food, or likes Kazakh customs—since some people who do none of these things are still Kazakhs. If you are born of Kazakh parents you certainly are a Kazakh, and remain so for the rest of your life, whatever changes may occur in your behavior. There is something special, some "Kazakhness," so to speak, that makes you behave the way you do, that creates a propensity to act like other Kazakhs, although that "something" is left undefined.[13]

Such representations of social groups as natural are found the world over. In many tribal societies, a widespread model construes lineages as based on the transmission of a special substance, metaphorically described as the permanent "bones" or "blood" or other substances embodied in transient individuals.[14] The caste ideology of classical India is another example. People belong to specific groups or *jāti*, each of which is traditionally attached to particular tasks or professions, for example, street sweepers or blacksmiths or royal functionaries or merchants, that are supposed to be so fundamentally different from each other that members of two distinct groups cannot generally share food or lodgings, let alone have sex. All these groups were (and in many places still are) ranked in strict order of purity— and the thought of contact with a member of an inferior caste is to many people disgusting. The traditional ideology made a clear connection between occupation and status. Tanners or undertakers were quite impure because of their contact with corpses. But this connection was not really considered the reason for someone's status. After all, Tanners or Butchers who stopped tanning or butchering, or indeed had never engaged in these activities, were considered just as polluting and disgusting as other members of their castes. In fact, in modern conditions most of the occupations traditionally associated with these castes have disappeared, but the castes themselves and their hierarchy are still present.[15]

Ideologies of natural differences are also invoked to justify ostracism against despised minorities. These may be members of culturally specific groups (Ainu in Japan, tribal people in India) or technical specialists (undertakers, blacksmiths, or potters in many societies in Africa and Asia). This form of social stratification is often accompanied by the notion that members of these groups are often thought to be naturally different from the rest of society. Finally, racialist ideologies may be the most salient manifestation of the belief in natural differences between groups. The notion of Jewish or black or white "blood" is among the metaphors that express the assumption of innate and ineradicable differences.[16]

In more general terms, people are essentialists about a social category when they assume that (a) all members share some special quality that is exclusive to the category and need not be defined, (b) possession of that special quality is a matter of biological descent, not historical accident or acquisition, and (c) that special quality is what makes them behave in particular ways.

Why is the ideology of essential natural differences so widespread and so compelling? Perhaps it is because our minds somehow mistake human groups for species.[17] Indeed, essentialist intuitions are very robust and explicit in representations of the natural world. Human minds intuitively construe animal species in terms of species-specific "causal essences." That is, their typical features and behavior are interpreted as consequences of the possession of an undefined yet causally relevant quality that is particular to each identified species. There is something special about giraffes, for instance, that they are born with, a giraffe-ness that makes them the way they are. This assumption appears early in child development, and it is implicit in our everyday biological knowledge. We acquire a lot of information about the various animals and plants we can observe, because we have an intuitive sense of how we can infer from instances to species. After observing a single animal behave, even young children know how to extend that information to all members of the species, and they use membership of a species to override perceptual appearances. Our biological essentialism is entrenched, and certainly adaptive.[18]

So it may be that social categories are construed as quasi-natural kinds because this kind of inference is already salient in human cognition. Also, some features of the representation of ethnies resemble some input conditions of the intuitive biology inference engine. According to Gil-White, humans process ethnic groups (and a few other related social categories) as if they were species because people apparently inherit ethnic identity from their parents, and ethnic groups (at least in some regions) do not intermarry.[19] In this view, people are somehow overextending their spontaneous essentialism whenever some cues match the input conditions of the essentialist inference engine. Our minds simply misconstrue ethnic categories for living kinds.[20] This would make sense, in a parsimonious way, of a great variety of recurrent cultural representations about groups.

The interpretation, however, is not completely satisfactory, at least not in this form. It implies that our intuitive biological systems make a mistake in considering social groups as species, simply because of the cues of endogamy and inheritance of group membership, in the same way as our visual system, for instance, can be fooled by two-dimensional perspective into seeing depth and volume where there aren't any. But that is not quite what happens. Endogamy is a prescriptive norm about marriage with strangers, not a fact about sex with them. In all societies where ethnic endogamy is enforced, people know perfectly well that it could be violated and would result in human offspring. You could have offspring with a Kazakh. Indeed, tribal warfare for much of human evolution included episodes of rape and abduction, suggesting that even when you call the other tribe subhumans or cockroaches, at least some of your brain systems take them to be human just like you, precisely as far as reproduction is concerned. In places where there are strict prohibitions against marrying outside the caste, people know perfectly well that sex across the caste line is possible and may result in viable offspring. That is indeed the whole point of enforcing strict caste prohibitions in the first place, and other similar laws against "miscegenation" between races.

So how can people hold beliefs about groups that seem compelling yet are partly inconsistent and include concepts that remain largely undefined? This is obviously not the case only in the domain of social essentialism, which serves here as an illustration of what happens in many other domains of cultural transmission. To understand this, we need yet more psychology.

Intuitions and Reflections about Other Groups

Here it may be of help to keep in mind that essentialist understandings of groups are not intuitive, in the sense of being spontaneous and effortless representations that just pop up, as it were, when we are faced with people of another group. Consider the range of automatic inferences and conjectures likely to come up when we encounter people of a different group. The conjectures will be about their intentions (Are they friendly? Do they want to trade? Is this an ambush?), about their capacities (Are there many of them? Do they look fierce, strong?), about their attractiveness, the oddity of their language or accent, and so on. But all these inferences or conjectures imply that we are dealing with human beings like us. We can measure their strength by looking at their arms, their aggressiveness by scrutinizing their faces, and so forth, because we use such cues to evaluate other people in our own group. So the intuitions, the mental representations delivered by automatic and largely unconscious mental systems, imply that the "Others" are of the same nature as "Us." There seems to be a clear discrepancy between these intuitions, on the one hand, and explicit statements, such as "different ethnic groups are like different species," on the other.

At this point I must explain in more detail the distinction, most clearly formulated by Dan Sperber, between intuitions or intuitive understandings, with their underlying cognitive machinery, on the one hand, and reflective information and beliefs, on the other. An intuition or intuitive understanding, for the sake of this argument, is simply the occurrence of some information that is potentially consciously accessible and directs the agent's expectations and behaviors, although the pathways that led to holding that information are not accessible to conscious inspection.[21] Consider for instance the following situations:

a) an infant expects a solid object on a collision course with a solid surface to bounce against it, not to fuse into it;[22]

b) after dissecting a crocodile and observing its innards, a person who is asked what is inside another crocodile spontaneously assumes that it must be the same stuff, but she is less confident if the second animal is a snake;[23] and

c) people primed with briefly flashed pictures of males from a minority group tend to misidentify subsequent pictures of tools as weapons, while they make the opposite mistake when primed with male faces from their own ethnic group.[24]

In each domain considered here, intuitive representations just pop up, so to speak, as a largely automatic and fast result of being presented with the relevant stimuli.

Reflective information, on the other hand, is information that has the effect of extending, making sense of, explaining, justifying, or communicating the contents of intuitive information. For instance, in the cases described so far, we can have the following reflective processes engaged:

a) people asked about the trajectories of objects explain them in terms of "impetus," "force," and "bouncing";

b) informants tell us that there is some unique quality in each animal that makes it a member of a species, and that it must be inherited—it cannot be acquired; and

c) people say that members of a particular ethnic group are lazy, aggressive, irresponsible, and so on, and that they are born that way—it is their nature.

This difference between these two kinds of mental representations should help us understand the cognitive processes most likely at play when people acquire information from others, in such a way that they build roughly similar representations, what we call "cultural" concepts or norms.

We can now return to the case of essentialism about social groups. Encounters with members of different groups are bound to trigger specific intuitive representations, as I described at some length in chapter 1 when discussing intergroup conflict. For instance, running into various members of the Poldovian people, and having information about the language they speak, what they eat, and what they wear, or whatever they do that is not familiar among one's own kin and acquaintances, we may have intuitions like these:

[1] Poldovians are not like us
[2] Poldovians are like each other
[3] Poldovians have common goals
[4] I cannot trust this person! She is a Poldovian

. . . which (may) trigger spontaneous reflective explanations like these:

[5] There must be something that makes Poldovians similar
[6] In some way, Poldovian newborns are already Poldovian

. . . and so forth. Now these reflective thoughts may occur to any individual, faced with differences in behavior, or any other differences, that seem to map onto social categories. But, obviously, such an individual lives in a social environment and acquires information about the relevant social categories from other individuals, information that may for instance contain statements like these:

[7] Once a Poldovian, always a Poldovian!

[8] Blood is thicker than water

[9] Different groups are like different species

. . . which, to some extent, carry implications that are very close to one's intuitions (for example [1–4]) and spontaneous reflections (for example [5–6]) and, to that extent, are likely to be attended to, included in people's store of plausible beliefs, and in some cases communicated to others (in a roughly similar form), creating the chains of transmission that constitute a tradition.

In cases like these, some specific representations are likely to become part of a tradition, because of the fit between the public statements (for example, that blood is thicker than water), on the one hand, and the contents of many individuals' intuitive and reflective representations, on the other. This talk of "fit" is of course much too vague. In the domain of social essentialism, the connection is that the explicit statements that people receive, and (often) repeat, provide a causal context for their prior intuitions. That is to say, reflective thoughts like "blood is thicker than water" provides what sounds like an explanation of prior intuitions. Other reflective representations, like "Once a Poldovian, always a Poldovian!" just seem to expand the scope of our intuitions, without really adding further explanation. So there may be many ways in which reflective representations are associated with our intuitions.

An important point here is that the reflective thoughts, either spontaneously activated in an individual mind or acquired from others, are not necessarily coherent, consistent, or satisfactory as explanations. That is certainly clear in the case of essentialist statements about ethnic groups or castes. The term "jāti" is used to qualify castes but also means "birth" or "species." It strongly suggests that members of different groups are as different as animals from different species. It may fit people's intuition that they would rather not mingle with members of other groups. But it does not really explain in what way groups are like living species. The same is true of other essentialist understandings, as I mentioned before. People may describe the other group as nonhuman or not fully human—that is indeed the most common form of xenophobia the world over—but such statements clash with many other intuitions (for instance, concerning these other people's thoughts and beliefs, their sexual attractiveness, and the like) that clearly indicate they are treated as humans.

Reflective representations can persist despite providing no explanations, poor explanations, or even incoherent explanations that conflict with our intuitions. This can happen, without individuals being irrational or confused, because, as Sperber pointed out, most reflective beliefs of this kind are metarepresentations, representations about representations.[25] For instance, the statement "Different groups are like different species" does not carry any clear implications about the way they are different, and is probably mentally represented as "In some way or other, it is true that 'different groups are like different species.'"

This lengthy description of the processes involved in transmitting a simple notion, like that of essentialized social groups, should suggest that cultural transmission is far from being the simple process we imagine, when we say that people just "absorb" the local "culture." In particular, it should also illustrate the fact that imitation is a terrible explanation for transmission. Imitation consists in copying surface features of observed behaviors. But that is not at all what happens in the transmission of essentialism for example. People can hear others saying that "Poldovians are different" or that "blood is thicker than water." They may sometimes repeat the words verbatim, which would be a case of imitation. But the fact that people are essentialist about social groups goes much further than that. It consists in people assuming an undefined quality that is present in all members of the category, and only those. It also consists in assuming that the internal quality can cause external behavior, but causation cannot go the other way around, so that external circumstances have no effect on essential qualities. But in the many societies where people have essentialist understandings of groups nobody ever uses such theoretical, indeed metaphysical, talk to describe groups. So the thoughts were produced not by imitating other people but by producing complex theoretical inferences (many of which remain unconscious) on the basis of other people's statements or behavior. In this case, the similarity in these thoughts is easy to explain, as an effect of the essentialist expectations we spontaneously entertain, particularly when thinking about living species. By transferring some (not all) assumptions and inference rules from our intuitive biology systems to our understanding of groups, we produce a partly coherent reflective representation of these groups. Because other people have the same form of biological essentialism as we do, their representations of groups ends up being very similar to ours. But, clearly, imitation is not what happened here.

Another general lesson from this particular example is that individual minds, instead of simply selecting information among what is offered by others, actively construct models that go far beyond the information given. I already mentioned that we can infer a correct melody from an out-of-tune rendition. Far from being exceptional, this process is central to the transmission of information, and we must describe it in more general terms.

Communication Requires Inferences

Human communication consists in reconstructing intentions. That is to say, when we communicate we do not upload mental representations from our minds to someone else's. What we do is produce some observable behavior, which together with masses of other information produces in the listener some representation of what we were intending to communicate. That view is both consistent with the evidence from pragmatics, the study of language use in conversation, and in direct conflict with a simpler, misleading but very persistent view, a code model of communication. In that code model, if we have a thought that we want to express, for example, that there is a large crocodile in the room, we use a series of symbols from our linguistic code, which produces the sentence "There is a large crocodile in the room!" Upon hearing that stream of sound symbols, the listener now has a new belief stored in her mind, namely, that there is a large crocodile in the room.

As linguists started to notice from the 1960s, there were many problems with that picture. One obvious problem is that most human communication occurs without any such close correspondence between what is said and what is meant. If I ask "Does she take care of her health?" and you reply "Well, she walks three miles on mountain trails every morning," you managed to communicate some information to the effect that she indeed takes care of her health. But it would be clearly absurd to conclude that the sentence "Well, she walks three miles on mountain trails every morning" is the code for communicating that particular fact. Indeed, that same sentence can be used to convey a very different meaning, for example, as an answer to the question, "Does she have a hard life, so young and so far away from school?" Linguists started revising the code model to make sense of such effects. As the philosopher Paul Grice pointed out, communication often seems to work as though speaker and listener had agreed on some tacit principles about the best way to convey information.[26] For instance, in actual conversations simply stating a true fact may also be highly misleading. Asked if he has children, King Lear might say "I have two daughters, Regan and Goneril"—but most people would find that utterance disingenuous, even though it is literally true. Not mentioning the third daughter violates Grice's principles of conversation.

In the end, pragmatics allows us to dispense with the encoding-decoding model altogether. This perspective was initiated by Grice, and developed by others, Deirdre Wilson and Dan Sperber in particular.[27] Rather than patch or repair the code model, pragmatics specialists argued that we should think of communication in a fundamentally different model, based on the perception of speakers' intentions. The main assumption of this "ostensive-inferential" model of communication is that the sender (that is, the speaker in the case of verbal exchange) produces external behavior (this is the ostensive part) designed to guide the receiver's cognitive process toward an interpretation of what he intended to communicate (this is the inferential part). As a reply to "Does she take care of her health?" the utterance "She walks three miles a day" suggests that the walks are to be considered as a clue that may lead to an answer to the original question. This is what Sperber and Wilson call the presumption of relevance. Producing an utterance signals that the sender expects the receiver to accept that what is produced is relevant to the topic at hand.

This is pertinent to cultural transmission, to those events that become parts of a chain of transmission and sometimes result in a tradition. Two characteristics are crucial here. One is that traditions are built by inference, that is, by the fact that minds go beyond the information given, to coin a phrase. The other is that inferences require background knowledge. Both are quite clear in the domain of everyday conversations. First, the thoughts that occur to an individual as the result of an utterance are generally not a direct translation of what was said. To understand how roughly similar representations could be held by Jack and Jill, as a result of Jill talking to Jack, the literal content of what she said to him is only the starting point. To proceed from that to his subsequent thoughts, we need to add a large amount of material that Jack spontaneously added to the utterance. Second, these inferences require that Jack mobilize previous information, notably knowledge stored in memory. For instance, "Well, she walks three miles on mountain trails every morning" could trigger the inference "She does take care of her health" or "Her life is hard, especially for a child," depending on the context. But each of these depends on activation of some additional information, to the effect that strenuous exercise is good for your health, in one case, and that it is rather rough on a child, in the other. There is no inferential work without such recruitment of stored information.

So if communication requires multiple inferences, and if these require prior knowledge, how do stability and change in traditions occur?

Attractors in Cognitive Space

Communication is an intrinsically entropic phenomenon. Communication follows unconstrained paths of inference. There is no way you can actually force some interpretation of your communicative behavior on others. So it would seem that inferences could go in all sorts of directions. At each point in a chain of communication, inferences could vary and create a proliferation of disparate representations.

That is indeed what happens—in most actual conversations. But once thousands of conversations are aggregated (the figure is a modest estimate, an indication of the order of magnitude of communication events in a small-scale community), there are also recurrent patterns. These are what evolutionary anthropologists call cognitive attractors.[28] They are attractors in the statistical sense, that is, patterns that recur in the context of otherwise random aggregations of specific events. To get a rough idea of what that means, consider pouring a liquid on a surface that is not perfectly level. The liquid will run from higher to lower points, in the process creating small puddles, places where the liquid is trapped, so to speak—these are called basins of attraction.

Now consider the abstract space of possible cultural concepts. When people entertain a particular concept, it is located at a particular place in that space. When they communicate with others, this may result in those people constructing a somewhat similar, somewhat different mental representation. This process, if it was entirely unconstrained, would result in concepts that occupy many different places in conceptual space, with equal probability. But cultural transmission seems to work like the distribution of liquid on an uneven surface. Some positions are more likely than others to be filled.

The previous six chapters provided many examples of such cultural attractors. The notion of a spirit or god, that is, a person with counterintuitive physical properties, occurs very often, while the idea of a plant with those same properties is very rare. Statues that listen to people are common, but not statues that grow as time passes. Many people construe social categories as groups whose members share some undefined, inherited essence—the notion that social groups are accidental collections of individuals is much rarer in cultural space. The notion that misfortune is caused by malevolent agents, often endowed with mysterious powers, is more widespread than the idea of misfortune as contingency, or indeed than any other possible interpretation of misfortune. The expectation that a marriage should be publicized, that it involves people other than bride and groom, is more widespread than the idea of a purely private contract. In the most various domains of cultural transmission we can observe that some concepts or norms are more likely than others to appear in human cultures.

Liquids gather in particular places because of gravitation. Cultural transmission ends up with recurrent mental representations, particular places in conceptual space, because of the inference systems that make some notions easier than others to acquire, entertain, and transmit.

Naturally, for any specific case of tradition, these very general attractors combine with other, local factors that make specific representations more likely than others. For instance, there may be a general trend, in human minds, toward interpreting misfortune as the effect of some agents' intentions. In a particular place, people may also have a tradition that describes the souls of the dead as errant and resentful ghosts. The effects of this local attractor would enhance those of the general disposition to think of misfortune as caused by agents. General attractors, manifest in many different cultures, exist because there are many similarities in all human minds. Local attractors can be observed because redundant or repeated transmission enhances the likelihood that particular details will be represented in roughly similar form by many minds in a community. Not to put too fine a point on it, all actual traditions manifest these two kinds of attractors, as Olivier Morin points out.[29]

At the beginning of this chapter I mentioned two very general questions about cultural transmission, namely, Why do people in a community (sometimes) happen to have roughly similar representations in particular domains? and Why are there recurrent features across very different places? The existence of attractors in conceptual space explains why these two are one and the same question. For instance, the availability of essentialism explains why people can infer the abstract notion of a shared essence from various particular statements about a social category, as I illustrated above in the case of the Poldovians. People in a group have the same intuitive expectations about species and their inherent qualities, which makes them converge on an essentialist interpretation of what they hear about particular groups in their environment. But that is true also across human groups. As essentialism about living things is general among human minds, we should not be surprised that its (partly coherent) extension to social groups is also found in so many societies. Similarity of representations inside a tradition, and recurrence of those representations across human societies, are caused by the same process.

Cognitive Tracks of Cultural Transmission

Attractors occur because communication requires inferences and access to prior knowledge. But, as I mentioned many times in the course of the previous chapters, mentally represented knowledge consists of many domain-specific stores of information, with their different input formats and inference rules. This would suggest that the emergence of attractors may take on very different forms and follow different processes, depending on the domain at hand. In other words, there could be very little that we can say about cultural transmission in general, because most of the action, so to speak, is a function of what inference systems are activated, and these systems have very different principles of operation.

If we consider distinct domains of behavior and representations that may show some similarity in a community, what we commonly call cultural material, it is true that the differences between domains are outstanding, and should be an essential part of any account of information transmission.

Consider for instance local ways of maintaining gaze, body posture, and appropriate distance in conversation. People in most of Africa would be embarrassed if you maintained eye contact with them—to them this would probably signal hostility—while most Europeans would be ill at ease if you did not. Such norms are sometimes made explicit (for example, "Don't point! Don't stare at people!") but also include many implicit expectations. What is the right distance to maintain during a conversation? The short reach that seems friendly to Americans would seem too great, possibly standoffish, to many Spaniards. There are very few systematic anthropological studies of these differences, even though they were identified as a puzzle for cultural transmission by no less an authority than Marcel Mauss, a founder of modern anthropology.[30] But what we know is that people in a particular community do have some parameters—they seem to agree, at least implicitly, on what is appropriate or too close for comfort—but no one is aware that they ever acquired that sense of appropriateness. It is only occasionally, when interacting with people from other places, that we realize that we have definite, though implicit, expectations in this domain, which we acquired without quite realizing it.[31]

As a contrasting case, consider the acquisition of stories. Whether we acquire stories in our tender years or later, this is an explicit process—that is, we are aware of the object that is learned, the plot, the characters, and the fact that all this is packaged as an object, for example, the story of the monkey that became a king. True, there are many aspects of story acquisition and retelling that remain tacit—notably the memory processes that make us forget or distort some aspects of what we heard—but the acquisition process itself is something we experience consciously. That way of acquiring something that is common to our community is, obviously, very different in its operation from the way we acquired the right distance to maintain between bodies.

Our knowledge of the natural world is another such domain, with yet another set of evolved dispositions and consequent constraints on traditions. Our knowledge of living species may vary a lot in its richness, from the atrophied competence that is typical of most modern societies, to the richly detailed botany and zoology acquired by members of small-scale communities with a simpler technology. Despite these differences, ethnobiology, as anthropologists call it, is based on similar principles the world over.[32] People use taxonomic frames to organize their biological classes, and the ranks used in biological knowledge are similar the world over.[33] Also, as I mentioned earlier, humans generally understand biological kinds in essentialist terms, that is, assume that each species or genus comes with a special internal substance that is inherited and produces the organism's external features and behaviors.[34]

So there are many different paths of communication, different tracks of cultural transmission. Even within a specific domain of behavior, some aspects of the common representations of the previous generation are acquired in systematically different ways. That is the case for language acquisition. Children learn the phonology, the lexicon, and the grammar of their language in different ways, operating distinct learning systems. One of those systems extracts a coherent phonology, a system of sounds from an environment that is, well, noisy, a process that starts even before birth.[35] Other systems provide heuristics for learning new words, producing plausible inferences about their meaning on the basis of previous knowledge, and systematic assumptions, such as the notion that most words correspond to unique mental concepts.[36] Other heuristics allow children to infer syntactic structures from the one-dimensional stream of speech.[37] The fact that different systems are involved, at different stages of development, explains why distinct aspects of language change at different paces and in different conditions. The lexicon can change fast, notably among a large population with many opportunities for communication. Phonology can also change, but much more slowly, and often does as a result either of contact between languages or of class differences indexed by different pronunciations.[38]

One could multiply the examples. All this suggests that there is no general formula for the emergence of cultural attractors, no general process that results in the transmission of cultural materials. Rather, there are parallel processes that affect different domains, different tracks of transmission—which we cannot describe unless we have a good understanding of the underlying, highly specific psychological systems involved.

This account of cultural transmission stands in contrast to the notion that inspired many meme-based models of cultural transmission, in which there is one general mechanism that accounts for the occurrence of similar representations in different minds. Considering the ways cultural material is acquired by human minds does not reveal the cultural acquisition process, in the sense of a general set of principles that would be true of many different domains. Rather, it shows that the transmission tracks differ a lot from domain to domain—and the clearest example is the transmission of technology.

The Ratchet: Toolmaking and Technology

Technology is, obviously, uniquely developed in the hominin lineage. There are many striking examples of apes and birds using twigs and rocks for ingenious instrumental purposes, but these only underscore the uniqueness of humans in this respect. Technology is also the human trait that has the most profound effect on our ecological niche, on the way humans modify environments in adaptive and maladaptive ways. Technology shows that human cultural transmission can be cumulative, adding material to previously transmitted information rather that replacing it, and in many cases adding material that makes our use of environments more efficient. The existence of technology raises the question of how humans managed to create ever more complex, and ever-improving, behaviors with an unchanged brain.

As technology changes a lot, it means that human minds must be able to extract the adequate information from what may be very different environments. Boyd and Richerson use the example of the construction of a kayak to emphasize that point. Kayaks are not simple or easy to make. (Incidentally, that is true of most "primitive" technology. Flint-knapping, or even the proper use of an atlatl, require quite a bit of knowledge and practice.) The proper use of proper materials, as well as the inevitable trade-offs between desirable features in a kayak, imply that individuals must acquire vast amounts of detailed technical knowledge by observing others and interacting with them. As Boyd and Richerson point out, the recipe for kayak construction is certainly not encoded, as such, in our genes.[39] That, of course, is true of most human behaviors, as I emphasized in the first pages of this book. So, in the same way as for other domains of behavior, the question is not, Does this specific behavior depend on large amounts of external information? (because they all do) but, What evolved capacities make it possible for individual minds to acquire that information?

In the same way as for language or music, highly specific evolved capacities are involved in the acquisition of technical skills. Indeed, these capacities appear early in cognitive development, both in the way young children learn to handle objects, on the basis of available external information, and in the way they organize their mental representations of different objects' functions. Even young children have a sophisticated understanding of functions. They categorize objects in terms of their intended function, the motivations involved in their creation, rather than their actual use, and this occurs in societies with very different technologies.[40] They spontaneously construe structural features of tools in terms of those intended functions.[41] But the crucial mechanism here is the way young children observe and learn from adult models' use of tools and machines. Young children are often said to "overimitate" technical gestures. In experimental settings, children confronted with a new set of behaviors associated with a new artifact, and with delivering a particular result, tend to reproduce all of the behaviors, including some that have no obvious causal connection to the goal and results. This tendency to associate results with all the detailed aspects of the behaviors that brought them about makes young children very different from chimpanzees, despite the latter's capacity for some tool use.[42] The term "overimitation" is misleading, as the children do not actually record and copy a series of gestures. For instance, they do not repeat the "irrelevant" actions if an adult has already performed them.[43] Although the phenomenon is well established, there is no consensus interpretation among developmental psychologists. One could see overimitation as an efficient learning strategy, balancing the low cost of including low-cost actions in one's performance against the possible loss of results. Or the tendency to overimitate could be linked to children's disposition to construe behaviors in a normative way, as the "proper" way to behave.[44]

These special capacities are involved in acquiring information about artifacts, tools, and instruments, rather than knowledge in general. They explain how the transmission of information makes it possible for people to build a kayak, on the basis of much observation, inference, and a lot of practice, as well as interaction with competent elders. Because of these cognitive capacities, there are certainly specific cognitive attractors in the transmission of technology, that is, combinations of ideas that are more likely than others to be included in stable technical traditions, although there is so far no systematic study of these effects of technical cognition on overall technological change.

Consideration of our evolved technical dispositions may also help us address the crucial question of technological change, namely, the possibility of accumulation and progress. In particular, we should try to understand how our technical dispositions allow the appearance of a ratchet effect, the process whereby techniques develop and then subsist rather than disappear, while other techniques build on the basis of previous ones rather than replacing them. Once a specific technique appears, it would seem, it is there to stay, which to a certain degree was the case for weapon manufacture, agriculture, metal smelting, and many other activities.

It is often tempting to see the emergence and cumulation of efficient techniques as a property of the human organism—speculating, for instance, that a special kind of imitation, or some radically new way of thinking, would explain the acceleration of progress in *Homo sapiens* after millennia and millennia of stagnation. But we should remember that technology is not, or not just, a matter of having clever individuals. It is also a population phenomenon, a matter of having individuals connected in the right way and motivated by the right incentives. In this respect, archaeology and history would suggest three main factors that explain the relatively late emergence of cumulative technology.

The first factor is division of labor, the necessary condition for creating techniques that, in their aggregate, go much further than the limits of one individual brain. Humans have known the minimal division of labor between the sexes for hundred of thousands of years. We know that even in technologically simple communities there is some division of tasks according to skills, some benefits from comparative advantage. We also know from the anthropological record, however, that this quickly reaches a demographic ceiling. The advantages of specialization become massive only when a large number of individuals are involved—what "large" means here is still an unresolved empirical question.

A second important factor, connected to the first one, is that communities can break the demographic limits by engaging in trade with surrounding groups. The explanation would be that trade, first, implies making the acquaintance of individuals with different traditions, and therefore allows what Matt Ridley calls "ideas having sex," that is, individuals being able to combine ingredients from different chains of transmission.[45] In other words, trade increases the number and diversity of chains of transmission, allowing the proliferation that sustains and changes traditions.[46]

A third factor, mostly relevant to the acceleration of technical change in large civilizations, is of course widespread literacy, as it multiplied the amount of technical information that could be passed on. Also, literacy allowed the appearance of blueprints, plans, tables of results, and other tools that gradually allowed the transformation of skilled craftsmen into early engineers.[47]

To sum up, having cumulative technology and then an accelerating technical progress does not require a special brain or a new brain, but it does require new conditions in which evolved minds can interact. The cognitive dispositions associated with toolmaking, and then recruited for the use of complex technology, are indeed special—they emerged in human evolution because of the fitness advantage provided by tools, which is why they are exquisitely appropriate to the understanding of tools.

Why Do People Believe in Culture? (And Nature?)

I started each chapter in the main part of this book with a specific question for the sciences of human societies. The substance of each chapter provided, not the definitive answer to any of these questions (obviously), but a large amount of information about the way the question can be addressed in an integrated account of societies, based on the findings and models of a whole variety of scientific research programs. I did not at any point refer to any demarcation or even distinction between "nature" and "nurture," or nature as opposed to culture. I trust the information provided in these chapters is enough to suggest that we can do a lot of scientific research without the hindrance of these confusing oppositions.

But these pointless distinctions are widespread, and despite the best efforts of proper scientists they are very much alive in the way research on human behavior is reported, with, for example, mention of particular behaviors being "hardwired" or "biological" or "innate," supposedly in contrast to others called "cultural" or "acquired." Some version of a nature-culture opposition, with the associated themes of universal versus variable, physiological versus mental, inevitable versus malleable, is among the crucial tenets of what John Tooby and Leda Cosmides called the Standard Social Science Model.[48] Such distinctions, one regrets to say, even disfigure otherwise well-informed discussions of human behavior, genes, evolution, and cultural differences. Why this extraordinary success?

Here is a speculative explanation. The nature-culture opposition may be one of those general cultural attractors I described earlier. Just like the notion of misfortune caused by malevolent agents, or that of social groups as having a special essence, the nature-culture opposition may be the probable result of some of our intuitive and reflective thoughts, in such a way that it will reappear in different guises, but with a similar set of principles, at different times and in different places, even though it is not entirely coherent and is in fact largely subverted by our best science.

As a first argument for this conjecture, note that an opposition of this kind crops up in the most different of human societies. The Greeks reflected on the opposition of *physis* and *nomos*, for instance, and had the most extraordinarily diverse ideas about how to delimit them, what aspects of human behavior could be attributed to one rather than the other.[49] Those were the scholarly, systematic reflections of literate scholars. But we also find some nature-culture opposition in many small-scale societies without specialized intellectuals. In most African societies, for example, there is a clear opposition between the worlds of the village, on the one hand, and bush or forest, on the other. One space is cultivated, governed by norms, while the other is wild, unpredictable, uncontrollable. Because there are no specialized intellectuals, these oppositions are not turned into an explicit and consistent theory of human behavior, but they work as implicit principles that organize people's notions of social life. The same can be said of most tribal societies of Asia or the Americas. Obviously, one should not ignore the many differences in such conceptions. As the anthropologist Philippe Descola has documented, people's conceptions of the distinction between nature and culture can give rise to very different forms of speculation, such as animism, in which people imagine that animals or spirits may have thought and intentionality, or totemism, which emphasizes the continuity between some human groups and some animal species.[50]

Despite the differences in the reflective and speculative paths that people follow in entertaining such questions, it may be that the questions themselves are grounded in highly similar intuitions. Here I am of course going much further than the evidence would warrant, but the starting point of this speculation is not really controversial. People the world over have similar mental systems, which trigger (roughly) similar kinds of intuitions about many aspects of the natural and social world. There is no evidence that people from different regions in the world, for instance, could confuse animate and inanimate objects, or that they could be thoroughly baffled by the notion of cooperation between nonkin, or that they would just not feel that killing those who helped you is immoral, or that they would not think of other people's behavior as governed by intentions and beliefs. Many other such intuitive systems guide our inferences and our acquisition of knowledge, as I mentioned in the previous chapters.

Now some of these systems might make a nature-culture opposition seem compelling. First, a form of mind-body dualism is almost inevitable in our reflections on behavior. It does not result from the adoption of philosophical theories about nature, since these are absent from most human societies. Instead, dualism is a straightforward result of the way our domain-specific inference systems are organized. Our intuitive psychology is a set of systems that produce descriptions and interpretations of behavior, notably of other people's behavior, in terms of invisible, indeed nonphysical, entities like thoughts and beliefs and intentions. From an early age, we construe them as nonmaterial, but we also assume that they have material effects, like, for example, making bodies move.[51] Our intuitions about bodies are grounded in specialized inferential systems about the physics of objects, while our intuitions about behaviors require appeals to these immaterial entities. Our evolved intuitive equipment does not provide causal bridges between these two systems, which is why all normal human beings are baffled when asked questions like, How did your intentions manage to move your arm? and even more puzzling, How did your intentions manage to move your arm in the particular way intended? So a notion of mental phenomena as separate from physical ones is, once formulated explicitly, quite compelling for human minds. Which is why some form of mind-body dualism is present in the most diverse human societies.[52]

Another set of intuitions may contribute to making some nature-culture opposition compelling. In all human communities, people have a conception of other groups as following norms different from "ours." Indeed, a distinction between one's own group as normal, central, and others as the aberrant case is the basis of much spontaneous ethnocentrism, found the world over. Also, as I mentioned in the previous chapter, children have no difficulty at all in acquiring the notion that social norms can be both different in distinct groups and objectively imperative, so to speak. So it may be easy to acquire a notion of "culture," as designating whatever is different between different human cultures. By contrast, we also have a large number of intuitive systems that apply to other people, regardless of their norms—we expect them to have, for example, preferences for well-being rather than pain, for fairness over exploitation, for alimentation and sexual pleasure over deprivation, and so forth. Which would provide some intuitive substance to what people would want to call "nature," thereby making the opposition between cultural and natural phenomena apparently substantial, despite its lack of actual reference, because it is coherent with our intuitive expectations.

Whether or not this speculative interpretation is valid, one thing is certain—the nature-culture opposition is not culturally successful because of its explanatory power, because of its success in making sense of human behavior. As we saw in previous chapters, human minds use information about their surroundings, including during childhood, to organize life strategies, for instance, to invest in the long-term or to reap the benefits of instant gratification. Systems in the mind make sense of group-level regularities to imagine some internal quality that is present in members of a social category. Other systems in the mind make it possible to entertain social interaction with agents that are not physically present, like dead people, mythical heroes, and gods and spirits. We also entertain partly coherent representations of mass-market economies, because we have templates for social exchange and fairness within small-scale communities. Among all these pieces of information that together produce human behaviors, which are nature and which are culture? No one knows, and it does not matter in the least—in fact no one could find out, because the separation is nowhere.

Thinking that humans have a unified domain of information called culture, and that this domain is separate from the realm of natural things, may be one among those Very Tempting Errors that the philosopher Dan Dennett warns scientists against, adding that one benefit of a liberal education is that we get to learn about all those mistakes, and what made them attractive, as an antidote against further forays down the same intellectual dead-ends.[53] Unfortunately, in the case of the imagined opposition between nature and culture, the Great Books are of no great help, as most of their authors were blindsided by the misleading intuitions I just described, as were most social scientists before progress in biology and psychology could provide an escape from this tempting error.

Philosophical Crumbs

A condition of scientific progress is to discard what philosophers call the manifest image, a picture of the world that seems both straightforward and self-evident.[54] In the physical world, that means abandoning our view of solid objects in a Euclidian space and replacing them with unintuitive notions of quantum objects. In the natural world, we had to jettison the notion of distinct natural species as essentially different, in order to think in terms of populations and of changing genotype frequencies.

As far as human social and cultural life is concerned, we need to exert a similar distancing effort, away from the manifest image of our social life. The many models and findings mentioned in the previous chapters show that the process is well under way, that we have the rudiments of some properly scientific accounts of (at least) some domains of human behavior. But the process is certainly effortful, many social scientists find it less than altogether compelling, and it may be difficult to explain to a general readership. Why is that?

There are many obstacles on the path to social science. One is that, as I discussed before, some ways of thinking about societies and cultural transmission are strongly influenced by our evolved dispositions. This is probably the case for the spontaneous, and highly contagious notion of culture versus nature. Also, many ways of thinking about human societies, including many efforts on the part of social scientists, are firmly entrenched in our folk sociology, whose expectations are probably an evolved system that makes social life possible, although it is a terrible tool for understanding social life.

Despite these obstacles, the convergence of research programs in many fields, as reported in the previous chapters, demonstrates that understanding human societies the scientific way is possible, even if we have only the fragments of such an understanding. This is the consequence of changes that occurred in many different disciplines, notably in cognitive psychology, neurosciences, evolutionary biology, and anthropology. The changes did not occur because scientists in these different fields adopted a new philosophy or a new, encompassing research program. Indeed, the social sciences in past centuries were hampered more than helped by manifestos and general philosophical pronouncements. For instance, in the early twentieth century Durkheim and Boas and other influential scholars made it an official tenet of the newly emerging social sciences that what happened in society had little or nothing to do with what biologists and psychologists were investigating. This kind of segregationist posturing persisted well into the century, making it very difficult for social scientists to realize how much they could gain by the integration of disciplines, by taking profit from the extraordinary developments of biology and cognitive sciences.

So, rather than a new philosophy, the scientific approach to human societies is grounded in a set of simple attitudes and healthy habits that are in fact rather natural to empirical scientists in other fields of inquiry. One of these is deliberate eclecticism, a decision to ignore disciplinary boundaries and traditions, so that evolutionary findings can inform history, economic models can be based on neurocognitive foundations, and cross-cultural comparisons on ecology and economics. The other habit is a healthy embrace of reductionism. For a long time, social scientists were horrified at the very notion of reduction, and they would clutch their pearls at the very thought of explaining social phenomena in terms of physiology, evolution, cognition, or ecology. The mere mention of psychological or evolutionary facts in descriptions of culture would, according to that academic version of the one-drop rule, irretrievably pollute the social scientific brew. But, in rejecting that form of reduction, social scientists were rejecting what is the common practice of most empirical scientists. Geologists do not ignore the findings and models of physics, they make constant use of them. The same goes for ecologists with biological findings, and for evolutionary biologists with molecular genetics. It was only recently that social scientists realized that these empirical disciplines were all actually making progress, and that may have to do with the systematic use of reduction in this sense, promising a vertical integration of different fields and disciplines.[55]

That integration is now happening. There is a great hope in these rudiments of a science that would follow the path originally traced by philosophers, historians, and moralists toward explaining the emergence of societies, a truly unique outcome of evolution by natural selection.

Notes

Introduction

1. Hinde, 1987; Rosenberg, 1980; E. O. Wilson, 1998.
2. Sanderson, 2014.
3. Seabright, 2012, pp. 15–61.
4. Foley, 1987.
5. Dawson, King, Bentley, and Ball, 2001; Gwinner, 1996.
6. Gallistel and King, 2011, pp. 2–25.
7. Sola and Tongiorgi, 1996.
8. Butterworth, 2001; Onishi and Baillargeon, 2005; Surian, Caldi, and Sperber, 2007; Woodward, 2003.
9. Harari, Gao, Kanwisher, Tenenbaum, and Ullman, 2016; Hooker et al., 2003; Pelphrey, Morris, and McCarthy, 2005.
10. Baron-Cohen, 1991, 1995.
11. Miklósi, Polgárdi, Topál, and Csányi, 1998; Povinelli and Eddy, 1996.
12. Sellars, 1963 [1991].
13. Cheraffedine et al., 2015.
14. B. L. Davis and MacNeilage, 1995; Werker and Tees, 1999.
15. Estes and Lew-Williams, 2015.
16. Pinker, 1984.
17. Hamlin, Wynn, and Bloom, 2007.
18. Blair et al., 1995.
19. Blair, 2007; Viding and Larsson, 2010.
20. Deardorff et al., 2010; Ellis et al., 2003; Ellis, Schlomer, Tilley, and Butler, 2012; Nettle, Coall, and Dickins, 2011; Quinlan, 2003.
21. Quinlan, Quinlan, and Flinn, 2003; Flinn, Ward, and Noone, 2005, p. 567; Jayakody and Kalil, 2002.
22. Edin and Kefalas, 2011.
23. Ellis et al., 2012; Mendle et al., 2009; Rowe, 2002; Waldron et al., 2007.
24. Del Giudice, 2009a.
25. H. C. Barrett, 2014, pp. 316–19; Boyer and Barrett, 2015; Sperber, 2002.
26. H. C. Barrett, 2014, pp. 26–27.

27. Gallistel and King, 2011, pp. 218–41.
28. Cosmides and Tooby, 1987; Tooby and Cosmides, 1995, 2005.
29. Maeterlinck, 1930, p. 52.
30. Leslie, Friedman, and German, 2004.
31. Dennett, 1987.
32. Rozin, Millman, and Nemeroff, 1986; Rozin and Royzman, 2001.
33. Carroll, Grenier, and Weatherbee, 2013.
34. Curtiss, Fromkin, Krashen, Rigler, and Rigler, 1974.
35. Kaufmann and Clément, 2007.
36. McCauley, 2011.

Chapter 1. What Is the Root of Group Conflict?

1. A. D. Smith, 1987.
2. E. Gellner, 1983.
3. Anderson, 1983; R. M. Smith, 2003; Wertsch, 2002.
4. Hobsbawm and Ranger, 1983.
5. Gat, 2013, pp. 67–131; A. D. Smith, 1987.
6. Gat, 2013.
7. Rotberg, 1999.
8. Brubaker, 2004.
9. Sorabji, 2006.
10. Brubaker, 2004, p. 7; Brubaker, Loveman, and Stamatov, 2004.
11. Brubaker, 2004, p. 167.
12. Ridley, 1996.
13. F. F. Chen and Kenrick, 2002; Gray, Mendes, and Denny-Brown, 2008; Krebs and Denton, 1997.
14. Kinzler, Shutts, Dejesus, and Spelke, 2009; Lev-Ari and Keysar, 2010; Nesdale and Rooney, 1996.
15. Boyer, Firat, and Van Leeuwen, 2015; De Dreu, Greer, Handgraaf, Shalvi, and Van Kleef, 2012; De Dreu, Greer, Van Kleef, Shalvi, and Handgraaf, 2011; Mendes, Blascovich, Lickel, and Hunter, 2002.
16. Norwich, 1989.
17. Armstrong, 1998.
18. Billig and Tajfel, 1973; Paladino and Castelli, 2008; Tajfel, 1970; Tajfel, Billig, and Bundy, 1971.
19. Rabbie, Schot, and Visser, 1989.
20. Rabbie et al., 1989.
21. Karp, Jin, Yamagishi, and Shinotsuka, 1993.

22. Kiyonari, Tanida, and Yamagishi, 2000; Yamagishi and Mifune, 2009.

23. Kurzban and Neuberg, 2005; Neuberg, Kenrick, and Schaller, 2010.

24. A. Y. Lee et al., 2010.

25. Fox, 2011, pp. 83–113; Pietraszewski, 2013.

26. Kurzban and Neuberg, 2005; Pietraszewski, 2013; Tooby and Cosmides, 2010.

27. Pietraszewski, 2013; Tooby and Cosmides, 2010.

28. Baron, 2001.

29. Kurzban, Tooby, and Cosmides, 2001; Pietraszewski, Cosmides, and Tooby, 2014.

30. Pietraszewski et al., 2014; Pietraszewski, Curry, Petersen, Cosmides, and Tooby, 2015.

31. Delton, Nemirow, Robertson, Cimino, and Cosmides, 2013.

32. Cimino and Delton, 2010; Delton and Cimino, 2010.

33. Dovidio, Gaertner, and Kawakami, 2003; Pettigrew and Tropp, 2008.

34. Bullock, 2013.

35. Hornsey, 2008.

36. Sidanius and Veniegas, 2000.

37. Sidanius and Pratto, 1999, pp. 52ff.

38. Payne, Lambert, and Jacoby, 2002; Sidanius and Veniegas, 2000.

39. Cosmides, Tooby, and Kurzban, 2003; Kurzban, Tooby, et al., 2001; Pietraszewski et al., 2014.

40. McGarty, Yzerbyt, and Spears, 2002.

41. Bradbury and Vehrencamp, 2000; Maynard Smith and Harper, 2003; Seyfarth and Cheney, 2003.

42. Scott-Phillips, 2008.

43. Mitchell, 1986; Searcy and Nowicki, 2010, pp. 3–6.

44. Jordan, 1979, pp. 75ff.

45. Gambetta, 2011.

46. Kuran, 1998.

47. Horowitz, 2001.

48. Horowitz, 2001, pp. 71–123.

49. Gat, 2006; Kalyvas, 2006.

50. English, 2003.

51. Taylor, 1999.

52. Dutton, 2007.

53. H. C. Barrett, 2005; Nell, 2006.

54. De Sales, 2003.

55. Luft, 2015.

56. Kalyvas, 2006.

57. Lakoff and Johnson, 1980.
58. Griskevicius et al., 2009.
59. Hobbes, 1651.
60. Rousseau, 1762.
61. Hrdy, 2009 pp. 27ff.
62. Chagnon, 1988; Daly and Wilson, 1988; M. Wilson and Daly, 1997.
63. LeBlanc and Register, 2003.
64. Gat, 2006; Horowitz, 2001.
65. LeBlanc and Register, 2003; Mueller, 2004.
66. Gat, 2006.
67. Wrangham and Peterson, 1997.
68. Gat, 2006 pp. 97ff.; Herz, 2003.
69. McDonald, Navarrete, and van Vugt, 2012.
70. Tooby and Cosmides, 1988, 2010.
71. McGarty et al., 2002.
72. Jussim, Crawford, and Rubinstein, 2015; Jussim, Harber, Crawford, Cain, and Cohen, 2005.
73. Gigerenzer, 2002; Gigerenzer and Hoffrage, 1995; Gigerenzer and Murray, 1987.
74. Bowen, 2006.
75. Putnam, 2000, 2007.
76. Dinesen and Sønderskov, 2012.
77. Williams and Mohammed, 2009.
78. Major, Mendes, and Dovidio, 2013.
79. Blascovich, Mendes, Hunter, Lickel, and Kowai-Bell, 2001; Page-Gould, Mendoza-Denton, and Tropp, 2008.
80. Boyer et al., 2015.
81. Alvarez and Levy, 2012; Bécares et al., 2012; Bosqui, Hoy, and Shannon, 2014; Das-Munshi et al., 2012; Das-Munshi, Becares, Dewey, Stansfeld, and Prince, 2010

Chapter 2. What Is Information For?

1. La Fontaine, 1998.
2. Bonhomme, 2012; Mather, 2005.
3. Douglas and Evans-Pritchard, 1970.
4. P. S. Boyer and Nissenbaum, 1974; Demos, 1982; Thomas, 1997.
5. Tooby and DeVore, 1987.
6. N. Carey, 2015.
7. Kant, 1781.
8. Mackay, 1841.

9. Blondel and Lévy-Bruhl, 1926.

10. S. Carey, 2009; Gallistel and Gelman, 2000; L. Hirschfeld and Gelman, 1994; Spelke and Kinzler, 2007.

11. Egyed, Király, and Gergely, 2013; Futó, Téglás, Csibra, and Gergely, 2010; Gergely, Egyed, and Király, 2007.

12. Kelemen, 2004; Kelemen and DiYanni, 2005.

13. Boyer and Barrett, 2005.

14. Asch, 1956.

15. Hyman, Husband, and Billings, 1995; Loftus, 1997.

16. Mercier, 2017.

17. Loftus, 1993, 2005.

18. Mercier, 2017.

19. Sperber et al., 2010.

20. Harris and Lane, 2014; Mascaro and Sperber, 2009.

21. Mercier and Sperber, 2011, 2017.

22. S. A. Thomas, 2007.

23. Bogart, Wagner, Galvan, and Banks, 2010; Klonoff and Landrine, 1999.

24. Allport and Postman, 1947.

25. Difonzo and Bordia, 2007; Whitson and Galinsky, 2008.

26. Baumeister, Bratslavsky, Finkenauer, and Vohs, 2001; Pratto and John, 1991.

27. Öhman, Flykt, and Esteves, 2001; Öhman and Mineka, 2001.

28. Boyer and Lienard, 2006; Neuberg, Kenrick, and Schaller, 2011; Woody and Szechtman, 2011.

29. Boyer and Bergstrom, 2011; Boyer and Lienard, 2006; Eilam, Izhar, and Mort, 2011; Öhman and Mineka, 2001; Rachman, 1977; Szechtman and Woody, 2004.

30. Blanchard, Griebel, and Blanchard, 2003; Woody and Szechtman, 2011.

31. Boyd and Richerson, 1985, pp. 213ff.

32. Fessler, Pisor, and Navarrete, 2014; Hilbig, 2009.

33. Boyer and Parren, 2015.

34. Brunvand, 1981; Eriksson and Coultas, 2014; Stubbersfield, Tehrani, and Flynn, 2014.

35. Lewandowsky, Ecker, Seifert, Schwarz, and Cook, 2012; Offit, 2011.

36. Festinger, Riecken, and Schachter, 1956.

37. Festinger, 1957.

38. Tedeschi, Schlenker, and Bonoma, 1971.

39. DeScioli and Kurzban, 2009, 2012; Tooby and Cosmides, 2010.

40. Baumard, André, and Sperber, 2013.

41. DeScioli and Kurzban, 2012, pp. 480–84; Tooby and Cosmides, 2010.

42. Tooby and Cosmides, 2010.

43. Douglas and Evans-Pritchard, 1970.
44. Hoffer, 1951.
45. English, 2003; Porta, 2008; A. G. Smith, 2008.
46. Cicero, 1923.
47. P. Boyer, 1990, pp. 61–78.
48. Leeson and Coyne, 2012.
49. Bernhardt and Allee, 1994; Huang, 1996; Katz, 2009.
50. Hutchins, 1980.
51. Merton, 1996.
52. Carruthers, Stich, and Siegal, 2002.
53. Douglas and Evans-Pritchard, 1970; Favret-Saada, 1980.
54. Bordia and Difonzo, 2004; Solove, 2007.
55. Bohner, Dykema-Engblade, Tindale, and Meisenhelder, 2008; Ross, Greene, and House, 1977; Wetzel and Walton, 1985.

Chapter 3. Why Are There Religions?

1. Harris, 1991; Roth, 2007.
2. Saler, Ziegler, and Moore, 1997.
3. Kant, 1790.
4. Ward, 1994, 1995.
5. J. L. Barrett, 2000; J. L. Barrett and Keil, 1996; P. Boyer, 1994.
6. P. Boyer and Barrett, 2005.
7. L. Hirschfeld and Gelman, 1994; Spelke, 2000.
8. J. L. Barrett, 1998; J. L. Barrett and Keil, 1996; J. L. Barrett and Nyhof, 2001; P. Boyer and Ramble, 2001; Gregory and Greenway, 2017.
9. Lloyd, 2007.
10. P. Boyer, 2000b.
11. M. Bloch and Parry, 1982.
12. Mallart Guimerà, 1981, 2003.
13. Stepanoff, 2014, pp. 113–51.
14. Needham, 1972.
15. Goody, 1986.
16. Kramer, 1961; Mann, 1955.
17. Whitehouse, 2000.
18. Demarest, 2004; Freidel, Schele, Parker, and Jay, 1993; Kramer, 1961; Sharer and Traxler, 2006.
19. Baumard and Boyer, 2013.
20. Beard, 1996; Cumont, 1910; Martin, 1987.

21. McCauley and Lawson, 2002; Whitehouse, 2000, 2004.
22. Whitehouse, 1995.
23. Evans-Pritchard, 1937, pp. 69ff.
24. D. N. Gellner, 1992.
25. J. L. Barrett, 1998, 2001; J. L. Barrett and Keil, 1996; J. L. Barrett, Richert, and Driesenga, 2001; Slone, 2004.
26. Jaspers, 1953.
27. Arnason, Eisenstadt, and Wittrock, 2005.
28. Eskildsen, 1998; Finn, 2009; Gombrich, 2006; Slingerland, 2007; Stark, 2003.
29. Musolino, 2015.
30. Gombrich, 2009; Stark, 2003.
31. Baumard and Boyer, 2013; Morris, 2006, 2013.
32. Grafen, 1990; Zahavi and Zahavi, 1997.
33. E. Bloch, 1985 [1921]; Scribner, 1990.
34. Eliade, 1959; Otto, 1920.
35. James, 1902.
36. Taves, 2009.
37. Luhrmann, 2012.
38. Luhrmann, 2012, pp. 132–56.
39. C. F. Davis, 1989.
40. P. Boyer, 2001, pp. 307–9; Sharf, 1998, 2000.
41. Whitehouse, 1992.
42. Irons, 2001.
43. Alcorta and Sosis, 2005; Bulbulia, 2004; Irons, 2001.
44. Bering, 2006; Norenzayan and Shariff, 2008; Shariff and Norenzayan, 2011.
45. P. Boyer, 2000a.
46. Stocking, 1984.
47. McCauley and Lawson, 1984.
48. M. Bloch, 2008.
49. Saler et al., 1997.
50. Kosmin, 2011; McCaffree, 2017.
51. Hanegraaff, 1998; Pike, 2012.
52. R. D. Putnam, 2010.
53. Gombrich and Obeyesekere, 1988.
54. Stewart, 2014.
55. Tambiah, 1992.
56. Bowen, 2007, 2010.
57. Bowen, 2010; Laurence and Vaïsse, 2006.
58. Bowen, 2012.

Chapter 4. What Is the Natural Family?

1. Malinowski, 1929.
2. Fox, 1967, p. 40.
3. Needham, 1971.
4. Jones, 2003.
5. Fortes, 1950, pp. 261ff.
6. Goody, 1990.
7. G. Childs, 2003; Goldstein, 1981.
8. Levine and Silk, 1997.
9. Goldstein, 1978; E. A. Smith, 1998.
10. Byrne and Whiten, 1988; Dunbar, 1993, 2003.
11. Wrangham, Jones, Laden, Pilbeam, and Conklin-Brittain, 1999.
12. Aiello and Wheeler, 1995.
13. Hrdy, 2009.
14. Hrdy, 1981, pp. 146ff.; Van Schaik and Van Hooff, 1983.
15. Campbell and Ellis, 2005; Fletcher, Simpson, Campbell, and Overall, 2015; Jankowiak and Fischer, 1992.
16. Abbott, 2011; Walker, Hill, Flinn, and Ellsworth, 2011.
17. Geary, 2005; Marlowe, 2000.
18. Fletcher et al., 2015; Geary, 2005; Gordon, Zagoory-Sharon, Leckman, and Feldman, 2010.
19. Washburn and Lancaster, 1968.
20. Hawkes and Bliege Bird, 2002; S. B. Hrdy, 2009, pp. 146ff.
21. Gurven and Hill, 2009.
22. Chapais, 2009; Gurven, 2004; Gurven and Hill, 2009.
23. Hrdy, 2009, pp. 151ff.; Mesnick, 1997.
24. Hrdy, 1977.
25. Arnold and Owens, 2002.
26. Anderson, Kaplan, Lam, et al., 1999; Anderson, Kaplan, and Lancaster, 1999.
27. Feinberg, Jones, Little, Burt, and Perrett, 2005; Licht, 1976; Puts, 2005; Ryan and Guerra, 2014.
28. Amundsen and Forsgren, 2001; Fink, Grammer, and Matts, 2006.
29. Symons, 1979, 1992.
30. Buss, 2000, 2003; Low, 2000; Seabright, 2012.
31. Langlois et al., 2000; Rhodes, Proffitt, Grady, and Sumich, 1998.
32. Badahdah and Tiemann, 2005; Chang, Wang, Shackelford, and Buss, 2011; Kamble, Shackelford, Pham, and Buss, 2014; Li, Bailey, Kenrick, and Linsenmeier, 2002.
33. Buss and Shackelford, 2008; Li et al., 2002.

34. Buss, 2003, pp. 75ff.

35. Cronin, 1991.

36. Daly and Wilson, 2001.

37. Buss and Schmitt, 1993; Kaplan and Gangestad, 2005.

38. Buss, 1989; Langlois et al., 2000; Schmitt, 2003; Sprecher, Sullivan, and Hatfield, 1994.

39. Buss and D. Schmitt, 1993; Miller and Todd, 1998.

40. Symons, 1992.

41. Fink and Penton-Voak, 2002; Perrett et al., 1999.

42. Goodenough and Heitman, 2014; Tooby, 1982.

43. D. Lieberman, Tooby, and Cosmides, 2007; Westermarck, 1921.

44. Wolf, 1995.

45. Kaplan and Gangestad, 2005; Roff, 2007; Stearns, 1992.

46. Griskevicius, Tybur, Delton, and Robertson, 2011; Hawkes, 2006; Kaplan and Gangestad, 2005.

47. Hill and Kaplan, 1999; Kaplan, Hill, Lancaster, and Hurtado, 2000.

48. Nettle, 2010; Nettle, Colléony, and Cockerill, 2011; Nettle et al., 2007.

49. Ellis, Figueredo, Brumbach, and Schlomer, 2009.

50. Belsky, Steinberg, and Draper, 1991; Del Giudice, 2009b; Rosenblum and Paully, 1984; Stearns, Allal, and Mace, 2008.

51. Del Giudice, Gangestad, and Kaplan, 2016; Ellis et al., 2009.

52. Elias et al., 2016; Le Roy Ladurie, 1975.

53. De Souza and Toombs, 2010.

54. Chapais, 2009, p. 161.

55. Becker, 1973, 1974.

56. Becker, 1981; E. A. Posner, 2000.

57. P. Boyer and Petersen, 2012.

58. Hannagan, 2008; Liesen, 2008; Rubin, 2002, p. 114.

59. R. B. Lee, 1979.

60. Gat, 2006, pp. 18ff.; LeBlanc and Register, 2003.

61. Alesina, 2013.

62. Smuts, 1995.

63. van Vugt, Cremer, and Janssen, 2007.

64. Baker et al., 2016.

65. Baumeister and Sommer, 1997; Gabriel and Gardner, 1999.

66. Geary, 1998, 2003.

67. Seabright, 2012, pp. 127ff.

68. van Vugt et al., 2007.

69. Human Rights Watch, 2008.

70. Alberts, Altmann, and Wilson, 1996; Komdeur, 2001.

71. Setchell, Charpentier, and Wickings, 2005.

72. Buss and Shackelford, 1997.

73. Gangestad, Garver-Apgar, Cousins, and Thornhill, 2014; Goetz and Romero, 2011; Haselton and Gangestad, 2006.

74. Goetz and Romero, 2011; Miner, Shackelford, and Starratt, 2009; M. Wilson and Daly, 1992, 1998.

75. M. Wilson and Daly, 1992, 1998.

76. Sokol, 2011.

77. Afkhami, 1995; Freedom House, 2014; Ghanim, 2009.

78. Pew Research Center, 2013.

79. El-Solh and Mabro, 1994; Mernissi, 1987; Peters, 1978; Pew Research Center, 2013.

80. Betzig, 1986.

81. Al-Ghanim, 2009; Freedom House, 2014; Khan, 2006.

82. Sell, 2011; Sell, Tooby, and Cosmides, 2009.

Chapter 5. How Can Societies Be Just?

1. Rousseau, 1984 [1755].

2. Hrdy, 2009.

3. Boyd and Richerson, 2006.

4. A. Smith, 1767.

5. Hamilton, 1963; Maynard Smith, 1964, 1982; E. O. Wilson, 1975.

6. Plott, 1974; V. L. Smith, 1976.

7. Boyd and Richerson, 1990; Boyd and Richerson, 2002; Boyd and Richerson, 2006.

8. Camerer, 2003; Fehr, Schmidt, Kolm, and Ythier, 2006; Gueth and van Damme, 1998; Levitt and J. List, 2007.

9. Ernst Fehr et al., 2006; Kurzban, McCabe, Smith, and Wilson, 2001.

10. Henrich, Fehr, et al., 2001.

11. Fehr and Gächter, 2002.

12. Boyd, Gintis, Bowles, and Richerson, 2003; de Quervain et al., 2004; Fowler, Johnson, and Smirnov, 2005.

13. Boyd and Richerson, 1992; Boyd and Richerson, 2006.

14. Boyd and Richerson, 1992; Boyd and Richerson, 2006; Turchin, 2007, p. 130.

15. Dubreuil, 2010.

16. Levitt and List, 2007; List, 2007.

17. Declerck, Kiyonari, and Boone, 2009; van Dijk and Wilke, 1997.

18. Kurzban, Descioli, and O'Brien, 2007.
19. Krasnow, Delton, Cosmides, and Tooby, 2016.
20. Burton-Chellew and West, 2013.
21. Gurven and Winking, 2008.
22. Baumard and Lienard, 2011; Price, 2005; Wiessner, 2005.
23. Bshary, 2002; Bshary and Grutter, 2005.
24. Adam, 2010; Bshary, 2002; Bshary and Grutter, 2005.
25. Ferriere, Bronstein, Rinaldi, Law, and Gauduchon, 2002.
26. André and Baumard, 2011.
27. Binmore, 2005, pp. 63–66.
28. Krasnow, Cosmides, Pedersen, and Tooby, 2012; Milinski, Semmann, and Krambeck, 2002; Noë and Hammerstein, 1994; Noë, van Schaik, and Van Hooff, 1991; Piazza and Bering, 2008.
29. Henrich et al., 2001.
30. Bshary and Bergmüller, 2008; Hagen and Hammerstein, 2006; Karp et al., 1993; Kiyonari et al., 2000.
31. Hagen and Hammerstein, 2006; Krasnow, Delton, Tooby, and Cosmides, 2013.
32. Delton, Krasnow, Cosmides, and Tooby, 2011.
33. Barclay, 2016.
34. R. D. Putnam, 2002.
35. Krasnow et al., 2012.
36. Baumard, Mascaro, and Chevallier, 2012; Chevallier et al., 2015; Lienard, Chevallier, Mascaro, Kiura, and Baumard, 2013.
37. André and Baumard, 2011; Baumard et al., 2013.
38. Gurven, 2004.
39. Sperber and Baumard, 2012.
40. Baumard et al., 2013; Baumard and Sheskin, 2015; Delton and Robertson, 2012, p. 52; Krasnow et al., 2016; Tomasello, 2009, pp. 52ff.
41. Bliege Bird and Bird, 1997; Gurven, 2004; Jaeggi and Van Schaik, 2011; H. Kaplan and Gurven, 2005.
42. Gurven, 2004.
43. Gurven, 2004; Gurven, Hill, Kaplan, Hurtado, and Lyles, 2000.
44. Dillian, 2010.
45. Renfrew, 1969.
46. Earle, 2002.
47. M. K. Chen, Lakshminarayanan, and Santos, 2006; Glimcher, 2009; Padoa-Schioppa and Assad, 2006; Santos and Platt, 2014.
48. M. K. Chen et al., 2006; Santos and Platt, 2014.
49. Brown, 1991; Heine, 1997.

50. Friedman, 2010; Friedman and Neary, 2008.

51. Noles and Keil, 2011.

52. P. Boyer, 2015.

53. Cosmides, 1989; Cosmides and Tooby, 1992, 2005.

54. Sugiyama, 1996.

55. Delton, Cosmides, Guemo, Robertson, and Tooby, 2012.

56. Guzman and Munger, 2014.

57. Hann and Hart, 2011; Humphrey and Hugh-Jones, 1992.

58. Hann and Hart, 2011.

59. McCabe and Smith, 2001.

60. Polanyi, 2001 [1957].

61. Tomasello, 2009.

62. S. D. Levitt and J. A. List, 2007.

63. Ostrom, 1990, 2005.

64. Munger, 2010; Ostrom, 2005.

65. A. Smith, 1776; Xenophon, 1960, 8.2.5.

66. Ricardo, 1817.

67. R. A. Posner, 1980, 2001.

68. Hitchner, 2005; Hopkins, 1980; Scheidel and Friesen, 2009; Ward-Perkins, 2005.

69. Fafchamps, 2016.

70. Greif, 1993.

71. Read, Reed, Ebeling, and Friedman, 2009.

72. Saad, 2012.

73. Caplan, 2006; KFF, 1996; Wood, 2002; Worstall, 2014.

74. Caplan, 2001, 2008.

75. Caplan, 2008.

76. Rubin, 2013.

77. Kipnis, 1997; Yan, 1996.

78. Tomasello, 2008, 2009.

79. Delton et al., 2011; Krasnow et al., 2013.

80. Guzman and Munger, 2014.

81. Nozick, 1974.

82. Rawls, 1971.

83. Sowell, 2007, pp. 187–222.

84. Roemer, 1996; Sen, 2009.

85. G. Clark, 2008; McCloskey, 2006; Ridley, 2010.

86. Ferreira et al., 2015; Landes, 1998; Morris, 2013.

87. Acemoglu, Johnson, and Robinson, 2002; Acemoglu and Robinson, 2012; Lal, 2010; McCloskey, 2006; Mokyr, 1992; Ridley, 2010.

Chapter 6. Can Human Minds Understand Societies?

1. Hirschfeld, 1994, 2013.
2. Ramble, 2008.
3. Ramble, 2008, p. 284.
4. Ramble, 2008, pp. 261–310.
5. Fried, 1967; Maryanski and Turner, 1992; Service, 1965.
6. Kelly, 1995.
7. Trigger, 2003.
8. Barreiro and Quintana-Murci, 2010; Deschamps et al., 2016; Pickrell et al., 2009.
9. Leonardi, Gerbault, Thomas, and Burger, 2012.
10. Seabright, 2010.
11. Anton, Potts, and Aiello, 2014; Dubreuil, 2010.
12. Kübler, Owenga, Reynolds, Rucina, and King, 2015.
13. Rousseau, 1984 [1755].
14. Ibn Khaldūn, 1958.
15. Chong, 1991, pp. 32ff.; Hardin, 1982; Medina, 2007; Olson, 1965, pp. 32–34.
16. Medina, 2007, p. 24.
17. Chong, 1991, p. 103; Hardin, 1995, pp. 50ff.; Medina, 2007, pp. 51ff.; Schelling, 1978, p. 101.
18. P. Boyer, 2008.
19. Somit and Peterson, 1997.
20. Rubin, 2002.
21. van Vugt, 2006.
22. King, Johnson, and van Vugt, 2009; van Vugt, 2006.
23. Nietzsche, 1882, § 13; 1980 [1901].
24. Betzig, 1986; Macfarlan, Walker, Flinn, and Chagnon, 2014.
25. Buss, 1989; Kamble et al., 2014; Rubin, 2002, pp. 114ff.
26. Boehm, 1999.
27. Clastres, 1989.
28. Graeber, 2007, pp. 303ff.
29. Kropotkin, 1902.
30. Herz, 2003; McDermott, 2004; Tetlock and Goldgeier, 2000.
31. See, e.g., Goodin, 1996; Sears, Huddy, and Jervis, 2003.
32. Sulikowski, 1993.
33. Bowden, 1979; Keesing, 1984.
34. Kirch, 2010, pp. 38ff.
35. Keesing, 1984.
36. Valeri, 1985, pp. 140ff.
37. Evans-Pritchard, 1962; Michelle Gilbert, 2008; Quigley, 2005.

38. Jowett, 2014; Kettell, 2013; Zivi, 2014.
39. Mallart Guimerà, 2003.
40. Jolly and Thomas, 1992; Keesing, 1993.
41. Pettit, 2003; Tuomela, 2013.
42. Gilbert, 1989; Sheehy, 2012.
43. Schelling, 1971.
44. E. Gellner, 1969.
45. Munger, 2015.
46. Nereid, 2011.
47. Kaiser, Jonides, and Alexander, 1986.
48. Baillargeon, Kotovsky, and Needham, 1995; Spelke, 1990.
49. Povinelli, 2003.
50. Talmy, 1988.
51. Talmy, 1988, 2000.
52. Lakoff and Johnson, 1980, p. 15.
53. Brown, 1991.
54. Havel, 1985, cited by Kuran, 1995.
55. Kuran, 1995, pp. 118–27.
56. Bruce, 2010, pp. 81ff.; D. Childs, 1996, pp. 83ff.
57. Bicchieri, 2006, pp. 183ff.
58. Ash, 2014.
59. Kuran, 1995, pp. 261–75.
60. Piaget, 1932.
61. Turiel, 1983.
62. Gabennesch, 1990.
63. Göckeritz, Schmidt, and Tomasello, 2014; Rakoczy and Schmidt, 2013; Rakoczy, Warneken, and Tomasello, 2008.
64. Lewis, 1969.
65. Bicchieri, 2006, pp. 11–28.
66. Horne, 2001.
67. Appiah, 2011.
68. Gates, 2001.
69. Xiaoxiaosheng, 1993.
70. Appiah, 2011; Gates, 2001.
71. Lakoff, 1987; Lakoff and Johnson, 1980.
72. Feigenson, 2011.
73. Pettit, 2003.
74. Bicchieri, 2006.
75. M. D. Lieberman, Schreiber, and Ochsner, 2003; Marcus, 2013; McDermott, 2011.

76. Sowell, 2007.

77. Haidt, 2013; Haidt and Joseph, 2004.

78. Haidt, 2013; Graham, Haidt, and Nosek, 2009; Haidt and Graham, 2007.

79. Hibbing, Smith, and Alford, 2013, pp. 121ff.; Oxley et al., 2008.

80. Hechter, 1987, pp. 115ff.

81. Gosden, 1961; Wilkinson, 1891.

82. Hechter, 1987.

83. Buchanan and Tullock, 2004 [1962].

84. Caplan, 2008; Sowell, 2011.

85. Ramble, 2008.

86. Mercier and Sperber, 2011.

Conclusion

1. Tooby and DeVore, 1987.

2. Morin, 2016.

3. Ginzburg, 1963, 2017.

4. Tarde, 1903.

5. Boyd and Richerson, 1985.

6. Cavalli-Sforza and Feldman, 1981; Lumsden and Wilson, 1981.

7. Dawkins, 1976, pp. 189ff.

8. Boyd and Richerson, 1985.

9. Richerson and Boyd, 2005, pp. 69ff.

10. Boyd and Richerson, 1985, p. 8.

11. Aunger, 2000; Sperber, 2000b.

12. P. Boyer, 1990; Rothbart and Taylor, 1990.

13. Gil-White, 2001.

14. M. Bloch, 1993; Daniel, 1984.

15. Dumont, 1970; Quigley, 1993.

16. Rothschild, 2001.

17. P. Boyer, 1990; Rothbart and Taylor, 1990.

18. Gelman, 1985, 2003.

19. Gil-White, 2001.

20. Gil-White, 2001.

21. Sperber, 1997.

22. Baillargeon et al., 1995; Spelke, 1990.

23. Gelman, Coley, and Gottfried, 1994.

24. Payne, 2001.

25. Sperber, 1997, 2000a.

26. Grice, 1991 [1967].
27. Sperber and Wilson, 1986, 1995.
28. Claidière, Scott-Phillips, and Sperber, 2014; Claidière and Sperber, 2007; Sperber and Claidière, 2006.
29. Morin, 2016, p. 130.
30. Mauss, 1973 [1937].
31. Sussman and Rosenfeld, 1982.
32. Atran and Medin, 1999.
33. Atran, 1990, 1995.
34. Gelman et al., 1994.
35. Kuhl et al., 2006; Moon, Lagercrantz, and Kuhl, 2013.
36. E. V. Clark, 1993; Markson and Bloom, 1997; Xu and Tenenbaum, 2007.
37. Pinker, 1989.
38. Hombert and Ohala, 1982; Labov, 1964.
39. Boyd and Richerson, 2005, pp. 159ff.
40. Asher and Nelson, 2008; H. C. Barrett, Laurence, and Margolis, 2008; Bloom, 1996.
41. Kelemen, Seston, and Georges, 2012.
42. Nagell, Olguin, and Tomasello, 1993; Whiten, Custance, Gomez, Teixidor, and Bard, 1996.
43. Kenward, Karlsson, and Persson, 2011.
44. Keupp, Behne, Zachow, Kasbohm, and Rakoczy, 2015.
45. Ridley, 2010.
46. Morin, 2016, pp. 125ff.
47. Goody, 1977, 1986.
48. Tooby and Cosmides, 1992.
49. Lloyd, 2007.
50. Descola, 2009, 2013.
51. German and Leslie, 2000; Leslie, 1987, 1994; Wellmann and Estes, 1986.
52. Astuti, 2001; M. Bloch, 1998; Bloom, 2007; Sarkissian et al., 2010.
53. Dennett, 2014, pp. 19–28.
54. Sellars, 1963 [1991].
55. Slingerland, 2008; Slingerland and Collard, 2012; E. O. Wilson, 1999.

Bibliography

Abbott, E. (2011). *A History of Marriage: From Same-Sex Unions to Private Vows and Common Law: The Surprising Diversity of a Tradition* (1st U.S. ed.). New York: Seven Stories Press.

Acemoglu, D., Johnson, S., and Robinson, J. A. (2002). *The Rise of Europe: Atlantic Trade, Institutional Change and Economic Growth.* Cambridge, Mass.: National Bureau of Economic Research.

Acemoglu, D., and Robinson, J. (2012). *Why Nations Fail: The Origins of Power, Prosperity, and Poverty* (1st ed.). New York: Crown Business.

Adam, T. C. (2010). "Competition Encourages Cooperation: Client Fish Receive Higher-Quality Service When Cleaner Fish Compete." *Animal Behaviour, 79* (6), 1183–89. doi: 10.1016/j.anbehav.2010.02.023

Afkhami, M. (1995). *Faith and Freedom: Women's Human Rights in the Muslim World.* London: I. B. Tauris.

Aiello, L. C., and Wheeler, P. (1995). "The Expensive-Tissue Hypothesis: The Brain and the Digestive System in Human and Primate Evolution." *Current Anthropology, 36* (2), 199–221. doi: 10.1086/204350

Alberts, S. C., Altmann, J., and Wilson, M. L. (1996). "Mate Guarding Constrains Foraging Activity of Male Baboons." *Animal Behaviour, 51* (6), 1269–77. doi: http://dx.doi.org/10.1006/anbe.1996.0131

Alcorta, C. S., and Sosis, R. (2005). "Ritual, Emotion, and Sacred Symbols: The Evolution of Religion as an Adaptive Complex." *Human Nature, 16* (4), 323–59. doi: 10.1007/s12110-005-1014-3

Alesina, A. (2013). "On the Origins of Gender Roles: Women and the Plough." *Quarterly Journal of Economics, 128* (2), 469–530. doi: 10.1093/qje /qjt005

Al-Ghanim, K. A. (2009). "Violence against Women in Qatari Society." *Journal of Middle East Women's Studies, 5* (1), 80–93.

Allport, G. W., and Postman, L. J. (1947). *The Psychology of Rumor.* New York: H. Holt.

Alvarez, K. J., and Levy, B. R. (2012). "Health Advantages of Ethnic Density for African American and Mexican American Elderly Individuals." *American Journal of Public Health, 102* (12), 2240–42.

Amundsen, T., and Forsgren, E. (2001). "Male Mate Choice Selects for Female Coloration in a Fish." *Proceedings of the National Academy of Sciences, 98* (23), 13155–60. doi: 10.1073/pnas.211439298

Anderson, B. R. (1983). *Imagined Communities: Reflections on the Origin and Spread of Nationalism.* London: Verso.

Anderson, K. G., Kaplan, H. S., Lam, D., & Lancaster, J. (1999). "Paternal Care by Genetic Fathers and Stepfathers II: Reports by Xhosa High School Students." *Evolution and Human Behavior*, 20 (6), 433–51.

Anderson, K. G., Kaplan, H. S., and Lancaster, J. (1999). "Paternal Care by Genetic Fathers and Stepfathers I: Reports from Albuquerque Men." *Evolution and Human Behavior, 20* (6), 405–31.

André, J.-B., and Baumard, N. (2011). "The Evolution of Fairness in a Biological Market." *Evolution, 65* (5), 1447–56. doi: 10.1111/j.1558–5646.2011.01232.x

Anton, S. C., Potts, R., and Aiello, L. C. (2014). "Human Evolution: Evolution of Early Homo: An Integrated Biological Perspective." *Science, 345* (6192), 1236828. doi: 10.1126/science.1236828

Appiah, K. A. (2011). *The Honor Code: How Moral Revolutions Happen.* New York: W. W. Norton.

Armstrong, G. (1998). *Football Hooligans: Knowing the Score.* New York: Berg.

Arnason, J., Eisenstadt, S., and Wittrock, B. (2005). *Axial Civilizations and World History.* Leiden: Brill Academic.

Arnold, K. E., and Owens, I. P. F. (2002). "Extra-Pair Paternity and Egg Dumping in Birds: Life History, Parental Care and the Risk of Retaliation." *Proceedings of the Royal Society B, 269,* 1263–69.

Asch, S. E. (1956). "Studies of Independence and Conformity: A Minority of One Against a Unanimous Majority." *Psychological Monographs, 70* (9), 1–70.

Ash, T. G. (2014). *The Magic Lantern: The Revolution of '89 Witnessed in Warsaw, Budapest, Berlin and Prague.* London: Atlantic Books.

Asher, Y. M., and Nelson, D. G. K. (2008). "Was It Designed to Do That? Children's Focus on Intended Function in Their Conceptualization of Artifacts." *Cognition, 106* (1), 474–83. doi: 10.1016/j.cognition.2007.01.007

Astuti, R. (2001). "Are We All Natural Dualists? A Cognitive Developmental Approach." *Journal of the Royal Anthropological Institute,* 7 (3), 429–47.

Atran, S. A. (1990). *Cognitive Foundations of Natural History: Towards an Anthropology of Science.* Cambridge: Cambridge University Press.

———. (1995). "Classifying Nature across Cultures." In E. E. Smith, D. N. Osherson, et al. (eds.), *Thinking: An Invitation to Cognitive Science,* vol. 3 (2nd ed.), pp. 131–74. Cambridge, Mass.: MIT Press.

Atran, S. A., and Medin, D. L. (eds.). (1999). *Folkbiology*. Cambridge, Mass.: MIT Press.

Aunger, R. (ed.). (2000). *Darwinizing Culture: The Status of Memetics as a Science*. Oxford: Oxford University Press.

Badahdah, A. M., and Tiemann, K. A. (2005). "Mate Selection Criteria among Muslims Living in America." *Evolution and Human Behavior, 26* (5), 432–40. doi: 10.1016/j.evolhumbehav.2004.12.005

Baillargeon, R., Kotovsky, L., and Needham, A. (1995). "The Acquisition of Physical Knowledge in Infancy." In D. Sperber, D. Premack, and A. James-Premack (eds.), *Causal Cognition: A Multidisciplinary Debate,* pp. 79–115. Oxford: Clarendon Press.

Baker, J. M., Liu, N., Cui, X., Vrticka, P., Saggar, M., Hosseini, S. M. H., and Reiss, A. L. (2016). "Sex Differences in Neural and Behavioral Signatures of Cooperation Revealed by Fnirs Hyperscanning." *Scientific Reports, 6,* 26492. doi: 10.1038/srep26492

Barclay, P. (2016). "Partner Choice versus Punishment in Human Prisoner's Dilemmas." *Evolution and Human Behavior, 37,* 263–71. doi: 10.1016/j .evolhumbehav.2015.12.004

Baron, J. (2001). "Confusion of Group Interest and Self-Interest in Parochial Cooperation on Behalf of a Group." *Journal of Conflict Resolution, 45* (3), 283–96.

Baron-Cohen, S. (1991). "Precursors to a Theory of Mind: Understanding Attention in Others." In A. Whiten (ed.), *Natural Theories of Mind,* pp. 233–51. Oxford: Blackwell.

———. (1995). *Mindblindness: An Essay on Autism and Theory of Mind*. Cambridge, Mass.: MIT Press.

Barreiro, L. B., and Quintana-Murci, L. (2010). "From Evolutionary Genetics to Human Immunology: How Selection Shapes Host Defence Genes." *Nat. Rev. Genet, 11* (1), 17–30. doi: http://www.nature.com/nrg/journal/v11/n1/ suppinfo/nrg2698_S1.html

Barrett, H. C. (2005). "Adaptations to Predators and Prey." In D. M. Buss (ed.), *The Handbook of Evolutionary Psychology,* pp. 200–223. Hoboken, N.J.: John Wiley and Sons.

———. (2014). *The Shape of Thought: How Mental Adaptations Evolve*. Oxford: Oxford University Press.

Barrett, H. C., Laurence, S., and Margolis, E. (2008). "Artifacts and Original Intent: A Cross-Cultural Perspective on the Design Stance." *Journal of Cognition and Culture, 8* (1–2), 1–22. doi: 10.1163/156770908x289189

Barrett, J. L. (1998). "Cognitive Constraints on Hindu Concepts of the Divine." *Journal for the Scientific Study of Religion, 37,* 608–19.

——. (2000). "Exploring the Natural Foundations of Religion." *Trends in Cognitive Sciences, 4* (1), 29–34.

——. (2001). "How Ordinary Cognition Informs Petitionary Prayer." *Journal of Cognition and Culture, 1* (3), 259–69.

Barrett, J. L., and Keil, F. C. (1996). "Conceptualizing a Nonnatural Entity: Anthropomorphism in God Concepts." *Cognitive Psychology, 31* (3), 219–47.

Barrett, J. L., and Nyhof, M. (2001). "Spreading Non-Natural Concepts: The Role of Intuitive Conceptual Structures in Memory and Transmission of Cultural Materials." *Journal of Cognition and Culture, 1* (1), 69–100.

Barrett, J. L., Richert, R. A., and Driesenga, A. (2001). "God's Beliefs versus Mother's: The Development of Nonhuman Agent Concepts." *Child Development, 72* (1), 50–65.

Baumard, N., André, J.-B., and Sperber, D. (2013). "A Mutualistic Approach to Morality: The Evolution of Fairness by Partner-Choice." *Behavioral and Brain Sciences, 36* (1), 59–78.

Baumard, N., and Boyer, P. (2013). "Explaining Moral Religions." *Trends in Cognitive Sciences, 17* (6), 272–80.

Baumard, N., and Lienard, P. (2011). "Second or Third Party Punishment? When Self-Interest Hides behind Apparent Functional Interventions." *Proceedings of the National Academy of Sciences of the United States of America, 108,* 39.

Baumard, N., Mascaro, O., and Chevallier, C. (2012). "Preschoolers Are Able to Take Merit into Account When Distributing Goods." *Developmental Psychology, 48* (2), 492–98. doi: 10.1037/a0026598

Baumard, N., and Sheskin, M. (2015). "Partner Choice and the Evolution of a Contractualist Morality." In J. Decety, T. Wheatley, J. Decety, and T. Wheatley (eds.), *The Moral Brain: A Multidisciplinary Perspective,* pp. 35–48. Cambridge, Mass.: MIT Press.

Baumeister, R. F., Bratslavsky, E., Finkenauer, C., and Vohs, K. D. (2001). "Bad Is Stronger Than Good." *Review of General Psychology, 5* (4), 323–70. doi: 10.1037/1089-2680.5.4.323

Baumeister, R. F., and Sommer, K. L. (1997). "What Do Men Want? Gender Differences and Two Spheres of Belongingness: Comment on Cross and Madson." *Psychological Bulletin, 122* (1), 38–44.

Beard, M. (1996). "The Roman and the Foreign: The Cult of the 'Great Mother' in Imperial Rome." In N. Thomas and C. Humphrey (eds.), *Shamanism, History and the State,* pp. 164–88. Ann Arbor: University of Michigan Press.

Bécares, L., Shaw, R., Nazroo, J., Stafford, M., Albor, C., Atkin, K., . . . Pickett, K. (2012). "Ethnic Density Effects on Physical Morbidity, Mortality, and Health Behaviors: A Systematic Review of the Literature." *American Journal of Public Health, 102* (12), e33–e66.

Becker, G. S. (1973). "A Theory of Marriage: Part I." *Journal of Political Economy,*
    81 (4), 813–46.
———. (1974). "A Theory of Marriage." In T. W. Schultz (ed.), *Economics of the*
    Family: Marriage, Children, and Human Capital, pp. 299–351. Chicago:
    University of Chicago Press.
———. (1981). *A Treatise on the Family.* Cambridge, Mass.: Harvard University
    Press.
Belsky, J., Steinberg, L., and Draper, P. (1991). "Childhood Experience,
    Interpersonal Development, and Reproductive Strategy: An Evolutionary
    Theory of Socialization." *Child Development, 62* (4), 647–70. doi:
    10.2307/1131166
Bering, J. M. (2006). "The Folk-Psychology of Souls." *Behavioral and Brain Sci-*
    ences, 29 (5), 453–62.
Bernhardt, K., Huang, P. C., and Allee, M. A. (1994). *Civil Law in Qing and Repub-*
    lican China. Stanford, Calif.: Stanford University Press.
Betzig, L. (1986). *Despotism and Differential Reproduction: A Darwinian View of*
    History. New York: Aldine.
Bicchieri, C. (2006). *The Grammar of Society: The Nature and Dynamics of Social*
    Norms. Cambridge: Cambridge University Press.
Billig, M., and Tajfel, H. (1973). "Social Categorization and Similarity in Intergroup
    Behavior." *European Journal of Social Psychology, 3,* 27–52.
Binmore, K. (2005). *Natural Justice.* New York: Oxford University Press.
Blair, R. J. R. (2007). "The Amygdala and Ventromedial Prefrontal Cortex in Mo-
    rality and Psychopathy." *Trends in Cognitive Sciences, 11* (9), 387–92.
Blair, R. J. R., Sellars, C., Strickland, I., Clark, F., Williams, A., Smith, M., and
    Jones, L. (1995). "Emotion Attributions in the Psychopath." *Personality and*
    Individual Differences, 19 (4), 431–37.
Blanchard, D. C., Griebel, G., and Blanchard, R. J. (2003). "Conditioning and Re-
    sidual Emotionality Effects of Predator Stimuli: Some Reflections on Stress
    and Emotion." *Progress in Neuro-Psychopharmacology and Biological Psy-*
    chiatry, 27 (8), 1177–85.
Blascovich, J., Mendes, W. B., Hunter, S. B., Lickel, B., and Kowai-Bell, N. (2001).
    "Perceiver Threat in Social Interactions with Stigmatized Others." *Journal*
    of Personality and Social Psychology, 80 (2), 253–67.
Bliege Bird, R. L., and Bird, D. W. (1997). "Delayed Reciprocity and Tolerated
    Theft: The Behavioral Ecology of Food-Sharing Strategies." *Current*
    Anthropology, 38 (1), 49–78.
Bloch, E. (1985 [1921]). *Thomas Münzer als Theologe der Revolution.* Berlin:
    Suhrkamp.

Bloch, M. (1993). "Domain-Specificity, Living Kinds and Symbolism." In P. Boyer (ed.), *Cognitive Aspects of Religious Symbolism,* pp. 111–20. Cambridge: Cambridge University Press.

———. (1998). *How We Think They Think: Anthropological Approaches to Cognition, Memory and Literacy.* Boulder, Colo.: Westview Press.

———. (2008). "Why Religion Is Nothing Special but Is Central." *Philosophical Transactions of the Royal Society of London, Series B, Biological Sciences, 363* (1499), 2055.

Bloch, M., and Parry, J. (1982). "Introduction: Death and the Regeneration of Life." In M. Bloch and J. Parry (eds.), *Death and the Regeneration of Life,* pp. 187–210. Cambridge: Cambridge University Press.

Blondel, C., and Lévy-Bruhl, L. (1926). *La mentalité primitive.* Paris: Stock.

Bloom, P. (1996). "Intention, History and Artifact Concepts." *Cognition, 60,* 1–29.

———. (2007). "Religion Is Natural." *Developmental Science, 10* (1), 147–51. doi: 10.1111/j.1467-7687.2007.00577.x

Boehm, C. (1999). *Hierarchy in the Forest: The Evolution of Egalitarian Behavior.* Cambridge, Mass.: Harvard University Press.

Bogart, L. M., Wagner, G., Galvan, F. H., and Banks, D. (2010). "Conspiracy Beliefs about HIV Are Related to Antiretroviral Treatment Nonadherence among African American Men with HIV." *JAIDS: Journal of Acquired Immune Deficiency Syndromes, 53* (5), 648–55.

Bohner, G., Dykema-Engblade, A., Tindale, R. S., and Meisenhelder, H. (2008). "Framing of Majority and Minority Source Information in Persuasion: When and How 'Consensus Implies Correctness.'" *Social Psychology, 39* (2), 108–16. doi: 10.1027/1864-9335.39.2.108

Bonhomme, J. (2012). "The Dangers of Anonymity: Witchcraft, Rumor, and Modernity in Africa." *HAU: Journal of Ethnographic Theory, 2* (2), 205–33.

Bordia, P., and Difonzo, N. (2004). "Problem Solving in Social Interactions on the Internet: Rumor as Social Cognition." *Social Psychology Quarterly, 67* (1), 33–49. doi: 10.1177/019027250406700105

Bosqui, T. J., Hoy, K., and Shannon, C. (2014). "A Systematic Review and Meta-Analysis of the Ethnic Density Effect in Psychotic Disorders." *Social Psychiatry and Psychiatric Epidemiology, 49* (4), 519–29.

Bowden, R. (1979). "Tapu and Mana: Ritual Authority and Political Power in Traditional Maori Society." *Journal of Pacific History, 14,* 50–61. doi: 10.1080/00223347908572364

Bowen, J. R. (2006). *Why the French Don't Like Headscarves: Islam, the State, and Public Space.* Princeton: Princeton University Press.

———. (2007). *Why the French Don't Like Headscarves: Islam, the State, and Public Space.* Princeton: Princeton University Press.

———. (2010). *Can Islam Be French? Pluralism and Pragmatism in a Secularist State.* Princeton: Princeton University Press.

———. (2012). *Blaming Islam.* Cambridge, Mass.: MIT Press.

Boyd, R., Gintis, H., Bowles, S., and Richerson, P. (2003). "The Evolution of Altruistic Punishment." *Proceedings of the National Academy of Sciences of the USA, 100* (6), 3531–35.

Boyd, R., and Richerson, P. J. (1985). *Culture and the Evolutionary Process.* Chicago: University of Chicago Press.

———. (1990). "Culture and Cooperation." In J. J. Mansbridge et al. (eds.), *Beyond Self-Interest,* pp. 111–32. Chicago: University of Chicago Press.

———. (1992). "Punishment Allows the Evolution of Cooperation (or Anything Else) in Sizable Groups." *Ethology and Sociobiology, 13,* 171–95.

———. (2002). "Group Beneficial Norms Can Spread Rapidly in a Structured Population." *Journal of Theoretical Biology, 215* (3), 287–96.

———. (2005). *The Origin and Evolution of Cultures.* Oxford: Oxford University Press.

———. (2006). "Solving the Puzzle of Human Cooperation." In S. C. Levinson and P. Jaisson (eds.), *Evolution and Culture,* pp. 105–32. Cambridge, Mass.: MIT Press.

Boyer, P. (1990). *Tradition as Truth and Communication: A Cognitive Description of Traditional Discourse.* Cambridge: Cambridge University Press.

———. (1994). "Cognitive Constraints on Cultural Representations: Natural Ontologies and Religious Ideas." In L. A. Hirschfeld and S. Gelman (eds.), *Mapping the Mind: Domain-Specificity in Culture and Cognition,* pp. 391–411. New York: Cambridge University Press.

———. (2000a). "Functional Origins of Religious Concepts: Conceptual and Strategic Selection in Evolved Minds." Malinowski Lecture 1999. *Journal of the Royal Anthropological Institute, 6,* 195–214.

———. (2000b). "Natural Epistemology or Evolved Metaphysics? Developmental Evidence for Early-Developed, Intuitive, Category-Specific, Incomplete, and Stubborn Metaphysical Presumptions." *Philosophical Psychology, 13* (3), 277–97.

———. (2001). *Religion Explained: Evolutionary Origins of Religious Thought.* New York: Basic Books.

———. (2008). "Evolutionary Economics of Mental Time Travel?" *Trends in Cognitive Sciences, 12* (6), 219–24.

———. (2015). "How Natural Selection Shapes Conceptual Structure: Human Intuitions and Concepts of Ownership." In E. Margolis and S. Laurence (eds.),

The Conceptual Mind: New Directions in the Study of Concepts, pp. 185–200. Cambridge, Mass.: MIT Press.

Boyer, P., and Barrett, H. C. (2005). "Domain Specificity and Intuitive Ontology." In D. M. Buss (ed.), *The Handbook of Evolutionary Psychology,* pp. 96–118. Hoboken, N.J.: John Wiley and Sons.

———. (2015). "Domain Specificity and Intuitive Ontologies." In D. M. Buss (ed.), *The Handbook of Evolutionary Psychology* (2nd ed.), pp. 161–80. Hoboken, N.J.: John Wiley and Sons.

Boyer, P., and Bergstrom, B. (2011). "Threat-Detection in Child Development: An Evolutionary Perspective." *Neuroscience and Biobehavioral Reviews, 35* (4), 1034–41.

Boyer, P., Firat, R., and Van Leeuwen, F. (2015). "Safety, Threat, and Stress in Intergroup Relations: A Coalitional Index Model." *Perspectives on Psychological Science, 10* (4), 434–50.

Boyer, P., and Lienard, P. (2006). "Why Ritualized Behavior? Precaution Systems and Action Parsing in Developmental, Pathological and Cultural Rituals." *Behavioral and Brain Sciences, 29* (6), 595–613.

Boyer, P., and Parren, N. (2015). "Threat-Related Information Suggests Competence: A Possible Factor in the Spread of Rumors." *PLoS One, 10* (6), e0128421. doi: 10.1371/journal.pone.0128421

Boyer, P., and Petersen, M. B. (2012). "The Naturalness of (Many) Social Institutions: Evolved Cognition as Their Foundation." *Journal of Institutional Economics, 8* (1), 1–25. doi: 10.1017/S1744137411000300

Boyer, P., and Ramble, C. (2001). "Cognitive Templates for Religious Concepts: Cross-Cultural Evidence for Recall of Counter-Intuitive Representations." *Cognitive Science, 25,* 535–64.

Boyer, P. S., and Nissenbaum, S. (1974). *Salem Possessed: The Social Origins of Witchcraft.* Cambridge: Mass.: Harvard University Press.

Bradbury, J. W., and Vehrencamp, S. L. (2000). "Economic Models of Animal Communication." *Animal Behaviour, 59* (2), 259–68.

Brown, D. E. (1991). *Human Universals.* New York: McGraw Hill.

Brubaker, R. (2004). *Ethnicity without Groups.* Cambridge, Mass.: Harvard University Press.

Brubaker, R., Loveman, M., and Stamatov, P. (2004). "Ethnicity as Cognition." *Theory and Society, 33* (1), 34.

Bruce, G. (2010). *The Firm: The Inside Story of the Stasi.* Oxford: Oxford University Press.

Brunvand, J. H. (1981). *The Vanishing Hitchhiker: American Urban Legends and Their Meanings.* New York: W. W. Norton.

Bshary, R. (2002). "Building Up Relationships in Asymmetric Co-Operation Games between the Cleaner Wrasse Labroides Dimidiatus and Client Reef Fish." *Behavioral Ecology and Sociobiology, 52* (5), 365–71. doi: 10.1007/s00265-002-0527-6

Bshary, R., and Bergmüller, R. (2008). "Distinguishing Four Fundamental Approaches to the Evolution of Helping." *Journal of Evolutionary Biology, 21* (2), 405–20. doi: 10.1111/j.1420-9101.2007.01482.x

Bshary, R., and Grutter, A. S. (2005). "Punishment and Partner Switching Cause Cooperative Behaviour in a Cleaning Mutualism." *Biology Letters, 1* (4), 396–99. doi: 10.1098/rsbl.2005.0344

Buchanan, J. M., and Tullock, G. (2004 [1962]). *The Calculus of Consent: Logical Foundations of Constitutional Democracy.* Indianapolis: Liberty Fund.

Bulbulia, J. (2004). "Religious Costs as Adaptations That Signal Altruistic Intention." *Evolution and Cognition,* 10(1), 19–42.

Bullock, D. (2013). "The Contact Hypothesis and Racial Diversity in the United States Military." Ph.D. dissertation, Texas Woman's University [dissertation number: AAI3550788]. Available from EBSCOhost psyh database.

Burton-Chellew, M. N., and West, S. A. (2013). "Prosocial Preferences Do Not Explain Human Cooperation in Public-Goods Games." *Proceedings of the National Academy of Sciences, 110* (1), 216–21. doi: www.pnas.org/cgi/doi/10.1073/pnas.1210960110

Buss, D. M. (1989). "Sex Differences in Human Mate Preferences: Evolutionary Hypotheses Tested in 37 Cultures." *Behavioral and Brain Sciences, 12,* 1–49.

———. (2000). *The Dangerous Passion: Why Jealousy Is as Necessary as Love and Sex.* New York: Free Press.

———. (2003). *The Evolution of Desire: Strategies of Human Mating.* New York: Basic Books.

Buss, D. M., and Schmitt, D. (1993). "Sexual Strategies Theory: An Evolutionary Perspective on Human Mating." *Psychological Review, 100,* 204–32.

Buss, D. M., and Shackelford, T. K. (1997). "From Vigilance to Violence: Mate Retention Tactics in Married Couples." *Journal of Personality and Social Psychology, 72* (2), 346–61. doi: 10.1037/0022-3514.72.2.346

———. (2008). "Attractive Women Want It All: Good Genes, Economic Investment, Parenting Proclivities, and Emotional Commitment." *Evolutionary Psychology, 6* (1), 134–46. doi: 10.1177/147470490800600116

Butterworth, G. (2001). "Joint Visual Attention in Infancy." In G. Bremner and A. Fogel (eds.), *Blackwell Handbook of Infant Development,* pp. 213–40. Malden, Mass.: Blackwell.

Byrne, R., and Whiten, A. (1988). *Machiavellian Intelligence: Social Expertise and the Evolution of Intellect in Monkeys, Apes, and Humans.* Oxford: Clarendon Press.

Camerer, C. (2003). *Behavioral Game Theory: Experiments in Strategic Interaction.* Princeton: Princeton University Press.

Campbell, L., and Ellis, B. J. (2005). "Commitment, Love, and Mate Retention." In D. M. Buss (ed.), *The Handbook of Evolutionary Psychology,* pp. 419–42. Hoboken, N.J.: John Wiley and Sons.

Caplan, B. (2001). "What Makes People Think Like Economists? Evidence on Economic Cognition from the 'Survey of Americans and Economists on the Economy.'" *Journal of Law and Economics, 44* (2), 395–426.

———. (2006). "How Do Voters Form Positive Economic Beliefs? Evidence from the Survey of Americans and Economists on the Economy." *Public Choice, 128* (3–4), 367–81. doi: 10.1007/s11127-006-9026-z

———. (2008). *The Myth of the Rational Voter: Why Democracies Choose Bad Policies* (new edition, with a new preface by the author). Princeton: Princeton University Press.

Carey, N. (2015). *Junk DNA: A Journey through the Dark Matter of the Genome.* London: Icon Books.

Carey, S. (2009). *The Origin of Concepts.* New York: Oxford University Press.

Carroll, S. B., Grenier, J. K., and Weatherbee, S. D. (2013). *From DNA to Diversity: Molecular Genetics and the Evolution of Animal Design.* New York: Wiley.

Carruthers, P., Stich, S., and Siegal, M. (eds.). (2002). *The Cognitive Basis of Science.* Cambridge: Cambridge University Press.

Cavalli-Sforza, L. L., and Feldman, M. W. (1981). *Cultural Transmission and Evolution: A Quantitative Approach.* Princeton: Princeton University Press.

Chagnon, N. A. (1988). "Life Histories, Blood Revenge, and Warfare in a Tribal Population." *Science, 239* (4843), 985–88.

Chang, L., Wang, Y., Shackelford, T. K., and Buss, D. M. (2011). "Chinese Mate Preferences: Cultural Evolution and Continuity across a Quarter of a Century." *Personality and Individual Differences, 50* (5), 678–83. doi: /10.1016/j.paid.2010.12.016

Chapais, B. (2009). *Primeval Kinship: How Pair-Bonding Gave Birth to Human Society.* Cambridge, Mass.: Harvard University Press.

Chen, F. F., and Kenrick, D. T. (2002). "Repulsion or Attraction? Group Membership and Assumed Attitude Similarity." *Journal of Personality and Social Psychology, 83* (1), 111–25.

Chen, M. K., Lakshminarayanan, V., and Santos, Laurie R. (2006). "How Basic Are Behavioral Biases? Evidence from Capuchin Monkey Trading Behavior." *Journal of Political Economy, 114* (3), 517–37. doi: 10.1086/503550

Chevallier, C., Xu, J., Adachi, K., van der Henst, J.-B., and Baumard, N. (2015). "Preschoolers' Understanding of Merit in Two Asian Societies." *PLoS One, 10* (5), e0114717. doi: 10.1371/journal.pone.0114717

Cheraffedine, R., Mercier, H., Clément, F., Kaufmann, L., Berchtold, A., Reboul, A., and Van der Henst, J.-B. (2015). "How Preschoolers Use Cues of Dominance to Make Sense of Their Social Environment." *Journal of Cognition and Development, 16* (4), 587–607. doi: 10.1080/15248372.2014.926269

Childs, D. (1996). *The Stasi: The East German Intelligence and Security Service.* New York: New York University Press.

Childs, G. (2003). "Polyandry and Population Growth in a Historical Tibetan Society." *History of the Family, 8,* 423–44.

Chong, D. (1991). *Collective Action and the Civil Rights Movement.* Chicago: University of Chicago Press.

Cicero, M. T. (1923). *De senectute, De amicitia, De divinatione.* Edited by W. A. Falconer. Loeb Classical Library 154. Cambridge, Mass.: Heinemann.

Cimino, A., and Delton, A. W. (2010). "On the Perception of Newcomers: Toward an Evolved Psychology of Intergenerational Coalitions." *Human Nature, 21* (2), 186–202. doi: 10.1007/s12110-010-9088-y

Claidière, N., Scott-Phillips, T. C., and Sperber, D. (2014). "How Darwinian Is Cultural Evolution?" *Philosophical Transactions of the Royal Society of London B: Biological Sciences, 369* (1642), 20130368. doi: 10.1098/rstb.2013.0368

Claidière, N., and Sperber, D. (2007). "The Role of Attraction in Cultural Evolution." *Journal of Cognition and Culture, 7* (1–2), 89–111.

Clark, E. V. (1993). *The Lexicon in Acquisition.* Cambridge: Cambridge University Press.

Clark, G. (2008). *A Farewell to Alms: A Brief Economic History of the World.* Princeton: Princeton University Press.

Clastres, P. (1989). *Society against the State: Essays in Political Anthropology.* New York: Zone Books.

Cosmides, L. (1989). "The Logic of Social Exchange: Has Natural Selection Shaped How Humans Reason? Studies with the Wason Selection Task." *Cognition, 31* (3), 187–276.

Cosmides, L., and Tooby, J. (1987). "From Evolution to Behavior: Evolutionary Psychology as the Missing Link." In J. Dupré (ed.), *The Latest on the Best: Essays on Evolution and Optimality,* pp. 297–323. Cambridge, Mass.: MIT Press.

———. (1992). "Cognitive Adaptations for Social Exchange." In J. H. Barkow, L. Cosmides, and J. Tooby (eds.), *The Adapted Mind: Evolutionary Psychology and the Generation of Culture*, pp. 163–228. New York: Oxford University Press.

——— (eds.). (2005). *Neurocognitive Adaptations Designed for Social Exchange.* Hoboken, N.J.: John Wiley and Sons.

Cosmides, L., Tooby, J., and Kurzban, R. (2003). "Perceptions of Race." *Trends in Cognitive Sciences, 7* (4), 173–79.

Cronin, H. (1991). *The Ant and the Peacock: Altruism and Sexual Selection from Darwin to Today.* Cambridge: Cambridge University Press.

Cumont, F. V. M. (1910). *The Mysteries of Mithra* (2nd ed.). Chicago: Open Court.

Curtiss, S., Fromkin, V., Krashen, S., Rigler, D., and Rigler, M. (1974). "The Linguistic Development of Genie." *Language, 50* (3), 528–54. doi: 10.2307/412222

Daly, M., and Wilson, M. (1988). *Homicide.* New York: Aldine.

———. (2001). "Risk-Taking, Intrasexual Competition, and Homicide." In J. A. French and A. C. Kamil (eds.), *Evolutionary Psychology and Motivation*, pp. 1–36. Lincoln: University of Nebraska Press.

Daniel, E. V. (1984). *Fluid Signs: Being a Person the Tamil Way.* Berkeley: University of California Press.

Das-Munshi, J., Bécares, L., Boydell, J. E., Dewey, M. E., Morgan, C., Stansfeld, S. A., and Prince, M. J. (2012). "Ethnic Density as a Buffer for Psychotic Experiences: Findings from a National Survey (EMPIRIC)." *British Journal of Psychiatry, 201* (4), 282–90.

Das-Munshi, J., Becares, L., Dewey, M. E., Stansfeld, S. A., and Prince, M. J. (2010). "Understanding the Effect of Ethnic Density on Mental Health: Multi-Level Investigation of Survey Data from England." *BMJ: British Medical Journal, 341* (7778), 1–9.

Davis, B. L., and MacNeilage, P. F. (1995). "The Articulatory Basis of Babbling." *Journal of Speech and Hearing Research, 38* (6), 1199–1211. doi: 10.1044/jshr.3806.1199

Davis, C. F. (1989). *The Evidential Force of Religious Experience.* Oxford: Clarendon Press.

Dawkins, R. (1976). *The Selfish Gene.* Oxford: Oxford University Press.

Dawson, A., King, V. M., Bentley, G. E., and Ball, G. F. (2001). "Photoperiodic Control of Seasonality in Birds." *Journal of Biological Rhythms, 16* (4), 365–80.

De Dreu, C. K. W., Greer, L. L., Handgraaf, M. J. J., Shalvi, S., and Van Kleef, G. A. (2012). "Oxytocin Modulates Selection of Allies in Intergroup Conflict." *Proceedings of the Royal Society B: Biological Sciences, 279* (1731), 1150–54.

De Dreu, C. K. W., Greer, L. L., Van Kleef, G. A., Shalvi, S., and Handgraaf, M. J. J. (2011). "Oxytocin Promotes Human Ethnocentrism." *Proceedings of the National Academy of Sciences, 108* (4), 1262–66.

de Quervain, D., Fischbacher, U., Treyer, V., Schellhammer, M., Schnyder, U., Buck, A., and Fehr, E. (2004). "The Neural Basis of Altruistic Punishment." *Science, 305,* 1254–58.

De Sales, A. (2003). "The Kham Magar Country: Between Ethnic Claims and Maoism." In D. N. Gellner (ed.), *Resistance and the State: Nepalese Experiences,* pp. 326–57. Oxford: Berghahn Books.

De Souza, M. J., and Toombs, R. J. (2010). "Amenorrhea Associated with the Female Athlete Triad: Etiology, Diagnosis, and Treatment." In F. N. Santoro and G. Neal-Perry (eds.), *Amenorrhea: A Case-Based, Clinical Guide,* pp. 101–25. Totowa, N.J.: Humana Press.

Deardorff, J., Ekwaru, J. P., Kushi, L. H., Ellis, B. J., Greenspan, L. C., Mirabedi, A., . . . Hiatt, R. A. (2010). "Father Absence, Body Mass Index, and Pubertal Timing in Girls: Differential Effects by Family Income and Ethnicity." *Journal of Adolescent Health, 48* (5), 441–47. doi: 10.1016/j.jadohealth.2010.07.032

Declerck, C. H., Kiyonari, T., and Boone, C. (2009). "Why Do Responders Reject Unequal Offers in the Ultimatum Game? An Experimental Study on the Role of Perceiving Interdependence." *Journal of Economic Psychology, 30* (3), 335–43. doi: 10.1016/j.joep.2009.03.002

Del Giudice, M. (2009a). "Human Reproductive Strategies: An Emerging Synthesis?" *Behavioral and Brain Sciences, 32* (1), 45–67. doi: 10.1017/S0140525X09000272

———. (2009b). "Sex, Attachment, and the Development of Reproductive Strategies." *Behavioral and Brain Sciences, 32* (1), 1–21. doi: 10.1017/S0140525X09000016

Del Giudice, M., Gangestad, S. W., and Kaplan, H. S. (2016). "Life History Theory and Evolutionary Psychology." In D. Buss (ed.), *The Handbook of Evolutionary Psychology,* vol. 1, pp. 88–114. New York: John Wiley and Sons.

Delton, A. W., and Cimino, A. (2010). "Exploring the Evolved Concept of NEW-COMER: Experimental Tests of a Cognitive Model." *Evolutionary Psychology, 8* (2), 317–35.

Delton, A. W., Cosmides, L., Guemo, M., Robertson, T. E., and Tooby, J. (2012). "The Psychosemantics of Free Riding: Dissecting the Architecture of a Moral Concept." *Journal of Personality and Social Psychology, 102* (6), 1252–70. doi: 10.1037/a0027026, 10.1037/a0027026.supp (Supplemental).

Delton, A. W., Krasnow, M. M., Cosmides, L., and Tooby, J. (2011). "Evolution of Direct Reciprocity under Uncertainty Can Explain Human Generosity in One-Shot Encounters." *PNAS Proceedings of the National Academy of*

Sciences of the United States of America, 108 (32), 13335–40. doi: 10.1073/pnas.1102131108

Delton, A. W., Nemirow, J., Robertson, T. E., Cimino, A., and Cosmides, L. (2013). "Merely Opting Out of a Public Good Is Moralized: An Error Management Approach to Cooperation." *Journal of Personality and Social Psychology, 105* (4), 621–38. doi: 10.1037/a0033495

Delton, A. W., and Robertson, T. E. (2012). "The Social Cognition of Social Foraging: Partner Selection by Underlying Valuation." *Evolution and Human Behavior, 33* (6), 715–25. doi: 10.1016/j.evolhumbehav.2012.05.007

Demarest, A. A. (2004). *Ancient Maya: The Rise and Fall of a Rainforest Civilization.* Cambridge: Cambridge University Press.

Demos, J. (1982). *Entertaining Satan: Witchcraft and the Culture of Early New England.* New York: Oxford University Press.

Dennett, D. C. (1987). *The Intentional Stance.* Cambridge, Mass.: MIT Press.

———. (2014). *Intuition Pumps and Other Tools for Thinking.* New York: W. W. Norton.

Deschamps, M., Laval, G., Fagny, M., Itan, Y., Abel, L., Casanova, J.-L., . . . Quintana-Murci, L. (2016). "Genomic Signatures of Selective Pressures and Introgression from Archaic Hominins at Human Innate Immunity Genes." *American Journal of Human Genetics, 98* (1), 5–21. doi: 10.1016/j.ajhg.2015.11.014

DeScioli, P., and Kurzban, R. (2009). "Mysteries of Morality." *Cognition, 112* (2), 281–99. doi: 10.1016/j.cognition.2009.05.008

———. (2012). "A Solution to the Mysteries of Morality." *Psychological Bulletin, 139* (2), 477–96. doi: 10.1037/a0029065

Descola, P. (2009). "Human Natures." *Social Anthropology, 17,* 145–57.

———. (2013). *Beyond Nature and Culture.* Chicago: University of Chicago Press.

Difonzo, N., and Bordia, P. (2007). *Rumor Psychology: Social and Organizational Approaches.* Washington, D.C.: American Psychological Association.

Dillian, C. D. W. C. L. (2010). "Trade and Exchange: Archaeological Studies from History and Prehistory" (document [dct]). New York: Springer.

Dinesen, P. T., and Sønderskov, K. M. (2012). "Trust in a Time of Increasing Diversity: On the Relationship between Ethnic Heterogeneity and Social Trust in Denmark from 1979 until Today." *Scandinavian Political Studies, 35* (4), 273–94. doi: 10.1111/j.1467-9477.2012.00289.x

Douglas, M., and Evans-Pritchard, E. E. (eds.). (1970). *Witchcraft Confessions and Accusations.* London: Tavistock.

Dovidio, J. F., Gaertner, S. L., and Kawakami, K. (2003). "Intergroup Contact: The Past, Present, and the Future." *Group Processes and Intergroup Relations, 6* (1), 5–20.

Dubreuil, B. (2010). *Human Evolution and the Origins of Hierarchies: The State of Nature.* New York: Cambridge University Press.

Dumont, L. (1970). *Homo Hierarchicus: An Essay on the Caste System.* Chicago: University of Chicago Press.

Dunbar, R. I. M. (1993). "Co-evolution of Neocortex Size, Group Size and Language in Humans." *Behavioral and Brain Sciences, 16* (4), 681–735.

———. (2003). "Evolution of the Social Brain." *Science, 302* (5648), 1160–61.

Dutton, D. G. (2007). *The Psychology of Genocide, Massacres, and Extreme Violence: Why "Normal" People Come to Commit Atrocities.* Westport, Conn.: Praeger Security International.

Earle, T. (2002). "Commodity Flows and the Evolution of Complex Societies." In J. Ensminger (ed.), *Theory in Economic Anthropology,* pp. 81–104. New York: Altamira Press.

Edin, K., and Kefalas, M. (2011). *Promises I Can Keep: Why Poor Women Put Motherhood before Marriage* (with a new preface). Berkeley: University of California Press.

Egyed, K., Király, I., and Gergely, G. (2013). "Communicating Shared Knowledge in Infancy." *Psychological Science, 24* (7), 1348–53.

Eilam, D., Izhar, R., and Mort, J. (2011). "Threat Detection: Behavioral Practices in Animals and Humans." *Neuroscience and Biobehavioral Reviews, 35* (4), 999–1006.

El-Solh, C. F., and Mabro, J. (eds.). (1994). *Muslim Women's Choices: Religious Belief and Social Reality.* Oxford: Berg.

Eliade, M. (1959). *The Sacred and the Profane: The Nature of Religion.* New York: Harcourt Brace Jovanovich.

Elias, S. G., van Noord, P. A. H., Peeters, P. H. M., den Tonkelaar, I., Kaaks, R., and Grobbee, D. E. (2016). "Menstruation during and after Caloric Restriction: The 1944–1945 Dutch Famine." *Fertility and Sterility, 88* (4), 1101–7. doi: 10.1016/j.fertnstert.2006.12.043

Ellis, B. J., Bates, J. E., Dodge, K. A., Fergusson, D. M., Horwood, L. J., Pettit, G. S., and Woodward, L. (2003). "Does Father Absence Place Daughters at Special Risk for Early Sexual Activity and Teenage Pregnancy?" *Child Development, 74* (3), 801–21.

Ellis, B. J., Figueredo, A. J., Brumbach, B. H., and Schlomer, G. L. (2009). "Fundamental Dimensions of Environmental Risk: The Impact of Harsh versus Unpredictable Environments on the Evolution and Development of Life History Strategies." *Human Nature, 20* (2), 204–68. doi: 10.1007/s12110-009-9063-7

Ellis, B. J., Schlomer, G. L., Tilley, E. H., and Butler, E. A. (2012). "Impact of Fathers on Risky Sexual Behavior in Daughters: A Genetically and

Environmentally Controlled Sibling Study." *Development and Psychopathology, 24* (1), 317–32. doi: 10.1017/S095457941100085X

English, R. (2003). *Armed Struggle: The History of the IRA.* Oxford: Oxford University Press.

Eriksson, K., and Coultas, J. C. (2014). "Corpses, Maggots, Poodles and Rats: Emotional Selection Operating in Three Phases of Cultural Transmission of Urban Legends." *Journal of Cognition and Culture, 14* (1–2), 1–26.

Eskildsen, S. (1998). *Asceticism in Early Taoist Religion.* Stony Brook, N.Y.: SUNY Press.

Estes, K. G., and Lew-Williams, C. (2015). "Listening through Voices: Infant Statistical Word Segmentation across Multiple Speakers." *Developmental Psychology, 51* (11), 1517–28. doi: 10.1037/a0039725

Evans-Pritchard, E. E. (1937). *Witchcraft, Oracles and Magic among the Azande.* Oxford: Clarendon Press.

———. (1962). "The Divine Kingship of the Shilluk of the Nilotic Sudan." In E. E. Evans-Pritchard (ed.), *Social Anthropology and Other Essays,* pp. 192–212. Glencoe, Ill.: Free Press.

Fafchamps, M. (2016). *Market Institutions in Sub-Saharan Africa: Theory and Evidence.* Cambridge, Mass.: MIT Press.

Favret-Saada, J. (1980). *Deadly Words: Witchcraft in the Bocage.* Cambridge: Cambridge University Press.

Fehr, E., and Gächter, S. (2002). "Altruistic Punishment in Humans." *Nature, 415,* 137–40.

Fehr, E., Schmidt, K. M., Kolm, S.-C., and Ythier, J. M. (2006). "The Economics of Fairness, Reciprocity and Altruism—Experimental Evidence and New Theories." In S.-C. Kolm and J. M. Ythier (eds.), *Handbook of the Economics of Giving, Altruism and Reciprocity,* vol. 1: *Foundations,* pp. 615–91. New York: Elsevier Science.

Feigenson, L. (2011). "Objects, Sets, and Ensembles." In S. Dehaene and E. Brannon (eds.), *Attention and Performance* (14), pp. 13–22. Oxford: Oxford University Press.

Feinberg, D. R., Jones, B., Little, A., Burt, D., and Perrett, D. (2005). "Manipulation of Fundamental and Formant Frequencies Influence Attractiveness of Human Male Voices." *Animal Behavior, 69,* 561–68.

Ferreira, F., Chen, S., Dabalen, A. L., Dikhanov, Y. M., Hamadeh, N., Jolliffe, D. M., . . . Yoshida, N. (2015). "A Global Count of the Extreme Poor in 2012: Data Issues, Methodology and Initial Results." World Bank Policy Working Paper, vol. 7432. Washington, D.C.: World Bank.

Ferriere, R., Bronstein, J. L., Rinaldi, S., Law, R., and Gauduchon, M. (2002). "Cheating and the Evolutionary Stability of Mutualisms." *Proceedings of the Royal Society of London B: Biological Sciences, 269* (1493), 773–80. doi: 10.1098/rspb.2001.1900

Fessler, D. M. T., Pisor, A., and Navarrete, C. D. (2014). "Negatively-Biased Credulity and the Cultural Evolution of Beliefs." *PLoS One, 9* (4), e95167.

Festinger, L. (1957). *A Theory of Cognitive Dissonance.* Stanford, Calif.: Stanford University Press.

Festinger, L., Riecken, H. W., and Schachter, S. (1956). *When Prophecy Fails.* Minneapolis: University of Minnesota Press.

Fink, B., Grammer, K., and Matts, P. (2006). "Visible Color Distribution Plays a Role in the Perception of Age, Attractiveness and Health in Female Faces." *Evolution and Human Behavior, 27,* 433–42.

Fink, B., and Penton-Voak, I. (2002). "Evolutionary Psychology of Facial Attractiveness." *Current Directions in Psychological Science, 11* (5), 154–58. doi: 10.1111/1467-8721.00190

Finn, R. D. (2009). *Asceticism in the Graeco-Roman World.* Cambridge: Cambridge University Press.

Fletcher, G. J. O., Simpson, J. A., Campbell, L., and Overall, N. C. (2015). "Pair-Bonding, Romantic Love, and Evolution: The Curious Case of Homo sapiens." *Perspectives on Psychological Science, 10* (1), 20–36. doi: 10.1177/1745691614561683

Flinn, M. V., Ward, C. V., and Noone, R. J. (2005). "Hormones and the Human Family." In D. M. Buss (ed.), *The Handbook of Evolutionary Psychology,* pp. 552–80. Hoboken, N.J.: John Wiley and Sons.

Foley, R. (1987). *Another Unique Species: Patterns in Human Evolutionary Ecology.* London: Longman Scientific and Technical.

Fortes, M. (1950). "Kinship and Marriage among the Ashanti." In A. R. Radcliffe-Brown and D. Forde (eds.), *African Systems of Kinship and Marriage,* pp. 252–84. London: Oxford University Press.

Fowler, J. H., Johnson, T., and Smirnov, O. (2005). "Egalitarian Motive and Altruistic Punishment." *Nature, 433,* e1–e2.

Fox, R. (1967). *Kinship and Marriage: An Anthropological Perspective.* Cambridge: Cambridge University Press.

———. (2011). *The Tribal Imagination: Civilization and the Savage Mind.* Cambridge, Mass.: Harvard University Press.

Freedom House. (2014). *Freedom in the World 2014: The Annual Survey of Political Rights and Civil Liberties.* Lanham, Md.: Rowman and Littlefield.

Freidel, D. A., Schele, L., Parker, J., and Jay, I. K. R. C. (1993). *Maya Cosmos: Three Thousand Years on the Shaman's Path.* New York: William Morrow.

Fried, M. H. (1967). *The Evolution of Political Society: An Essay in Political Anthropology*. New York: Random House.

Friedman, O. (2010). "Necessary for Possession: How People Reason about the Acquisition of Ownership." *Personality and Social Psychology Bulletin, 36* (9), 1161–69. doi: 10.1177/0146167210378513

Friedman, O., and Neary, K. R. (2008). "Determining Who Owns What: Do Children Infer Ownership from First Possession?" *Cognition, 107* (3), 829–49. doi: 10.1016/j.cognition.2007.12.002

Futó, J., Téglás, E., Csibra, G., and Gergely, G. (2010). "Communicative Function Demonstration Induces Kind-Based Artifact Representation in Preverbal Infants." *Cognition, 117* (1), 1–8. doi: 10.1016/j.cognition.2010.06.003

Gabennesch, H. (1990). "The Perception of Social Conventionality by Children and Adults." *Child Development, 61* (6), 2047–59. doi: 10.2307/1130858

Gabriel, S., and Gardner, W. L. (1999). "Are There 'His' and 'Hers' Types of Interdependence? The Implications of Gender Differences in Collective versus Relational Interdependence for Affect, Behavior, and Cognition." *Journal of Personality and Social Psychology, 77* (3), 642–55.

Gallistel, C. R., and Gelman, R. (2000). "Non-verbal Numerical Cognition: From Reals to Integers." *Trends in Cognitive Sciences, 4,* 59–65.

Gallistel, C. R., and King, A. P. (2011). *Memory and the Computational Brain: Why Cognitive Science Will Transform Neuroscience*. New York: Wiley.

Gambetta, D. (2011). *Codes of the Underworld: How Criminals Communicate*. Princeton: Princeton University Press.

Gangestad, S. W., Garver-Apgar, C. E., Cousins, A. J., and Thornhill, R. (2014). "Intersexual Conflict across Women's Ovulatory Cycle." *Evolution and Human Behavior, 35* (4), 302–8. doi: 10.1016/j.evolhumbehav.2014.02.012

Gat, A. (2006). *War in Human Civilization*. New York: Oxford University Press.

———. (2013). *Nations: The Long History and Deep Roots of Political Ethnicity and Nationalism*. Cambridge: Cambridge University Press.

Gates, H. (2001). "Footloose in Fujian: Economic Correlates of Footbinding." *Comparative Studies in Society and History, 43* (1), 130–48. doi: 10.2307/2696625

Geary, D. C. (1998). *Male, Female: The Evolution of Human Sex Differences*. Washington, D.C.: American Psychological Association.

———. (2003). "Evolution and Development of Boys' Social Behavior." *Developmental Review, 23,* 444–70.

———. (2005). "Evolution of Paternal Investment." In D. M. Buss (ed.), *The Handbook of Evolutionary Psychology*, pp. 483–505. Hoboken, N.J.: John Wiley and Sons.

Gellner, D. N. (1992). *Monk, Householder, and Tantric Priest: Newar Buddhism and Its Hierarchy of Ritual.* Cambridge: Cambridge University Press.

Gellner, E. (1969). *Saints of the Atlas.* London: Weidenfeld and Nicolson.

———. (1983). *Nations and Nationalism.* Oxford: Blackwell.

Gelman, S. A. (1985). *Children's Inductive Inferences from Natural Kind and Artifact Categories.* Stanford, Calif.: Stanford University Press.

———. (2003). *The Essential Child: Origins of Essentialism in Everyday Thought.* New York: Oxford University Press.

Gelman, S. A., Coley, J. D., and Gottfried, G. M. (1994). "Essentialist Beliefs in Children: The Acquisition of Concepts and Theories." In L. A. Hirschfeld and S. A. Gelman (eds.), *Mapping the Mind: Domain Specificity in Cognition and Culture,* pp. 341–65. New York: Cambridge University Press.

Gergely, G., Egyed, K., and Király, I. (2007). "On Pedagogy." *Developmental Science, 10* (1), 139–46.

German, T. P., and Leslie, A. M. (2000). "Attending to and Learning about Mental States." In P. Mitchell, K. J. Riggs, et al. (eds.), *Children's Reasoning and the Mind,* pp. 229–52. Hove, U.K.: Psychology Press/Taylor and Francis.

Ghanim, D. (2009). *Gender and Violence in the Middle East.* London: Praeger.

Gigerenzer, G. (2002). *Adaptive Thinking: Rationality in the Real World.* New York: Oxford University Press.

Gigerenzer, G., and Hoffrage, U. (1995). "How to Improve Bayesian Reasoning without Instruction: Frequency Formats." *Psychological Review, 102,* 684–704.

Gigerenzer, G., and Murray, D. J. (1987). *Cognition as Intuitive Statistics.* Hillsdale, N.J.: L. Erlbaum.

Gil-White, F. (2001). "Are Ethnic Groups Biological 'Species' to the Human Brain? Essentialism in Our Cognition of Some Social Categories." *Current Anthropology, 42* (4), 515–54. doi: 10.1086/321802

Gilbert, M. (1989). *On Social Facts.* London: Routledge.

———. (2008). "The Sacralized Body of the Akwapim King." In N. Brisch (ed.), *Religion and Power: Divine Kingship in the Ancient World and Beyond,* pp. 171–90. Chicago: Oriental Institute of the University of Chicago.

Ginzburg, N. (1963). *Lessico famigliare.* Turin: Einaudi.

———. (2017). *Family Lexicon.* Trans. J. McPhee. New York: New York Review Books.

Glimcher, P. W. (2009). "Neuroeconomics and the Study of Valuation." In M. S. Gazzaniga, E. Bizzi, L. M. Chalupa, S. T. Grafton, T. F. Heatherton, C. Koch, J. E. LeDoux, S. J. Luck, G. R. Mangan, J. A. Movshon, H. Neville, E. A. Phelps, P. Rakic, D. L. Schacter, M. Sur, B. A. Wandell, M. S. Gazzaniga, E. Bizzi, L. M. Chalupa, S. T. Grafton, T. F. Heatherton, C. Koch,

J. E. LeDoux, S. J. Luck, G. R. Mangan, J. A. Movshon, H. Neville, E. A. Phelps, P. Rakic, D. L. Schacter, M. Sur, and B. A. Wandell (eds.), *The Cognitive Neurosciences* (4th ed.), pp. 1085–92. Cambridge, Mass.: MIT Press.

Göckeritz, S., Schmidt, M. F. H., and Tomasello, M. (2014). "Young Children's Creation and Transmission of Social Norms." *Cognitive Development, 30,* 81–95. doi: 10.1016/j.cogdev.2014.01.003

Goetz, A. T., and Romero, G. A. (2011). "Family Violence: How Paternity Uncertainty Raises the Stakes." In C. Salmon, T. K. Shackleford, C. Salmon, and T. K. Shackleford (eds.), *The Oxford Handbook of Evolutionary Family Psychology,* pp. 169–80. New York: Oxford University Press.

Goldstein, M. C. (1978). "Pahari and Tibetan Polyandry Revisited." *Ethnology, 17,* 325–37.

———. (1981). "New Perspectives on Tibetan Fertility and Population Decline." *American Ethnologist, 8* (4), 721–38.

Gombrich, R. F. (2006). *Theravada Buddhism: A Social History from Ancient Benares to Modern Colombo.* London: Routledge.

———. (2009). *What the Buddha Thought.* London: Equinox.

Gombrich, R. F., and Obeyesekere, G. (1988). *Buddhism Transformed: Religious Change in Sri Lanka.* Princeton: Princeton University Press.

Goodenough, U., and Heitman, J. (2014). "Origins of Eukaryotic Sexual Reproduction." *Cold Spring Harbor Perspectives in Biology, 6* (3). doi: 10.1101/cshperspect.a016154

Goodin, R. E. K. H.-D. (1996). *A New Handbook of Political Science.* Oxford: Oxford University Press.

Goody, J. (1977). *The Domestication of the Savage Mind.* Cambridge: Cambridge University Press.

———. (1986). *The Logic of Writing and the Organization of Society.* Cambridge: Cambridge University Press.

———. (1990). *The Oriental, the Ancient and the Primitive: Systems of Marriage and the Family in the Pre-Industrial Societies of Eurasia.* Cambridge: Cambridge University Press.

Gordon, I., Zagoory-Sharon, O., Leckman, J. F., and Feldman, R. (2010). "Prolactin, Oxytocin, and the Development of Paternal Behavior across the First Six Months of Fatherhood." *Hormones and Behavior, 58* (3), 513–18. doi: 10.1016/j.yhbeh.2010.04.007

Gosden, P. (1961). *The Friendly Societies in England, 1815–1875.* Manchester: Manchester University Press.

Graeber, D. (2007). *Possibilities: Essays on Hierarchy, Rebellion and Desire.* Chico, Calif.: AK Press.

Grafen, A. (1990). "Biological Signals as Handicaps." *Journal of Theoretical Biology, 144* (4), 517–46.

Graham, J., Haidt, J., and Nosek, B. A. (2009). "Liberals and Conservatives Rely on Different Sets of Moral Foundations." *Journal of Personality and Social Psychology, 96* (5), 1029–46.

Gray, H. M., Mendes, W. B., and Denny-Brown, C. (2008). "An In-Group Advantage in Detecting Intergroup Anxiety." *Psychological Science, 19* (12), 1233–37. doi: 10.1111/j.1467-9280.2008.02230.x

Gregory, J. P., and Greenway, T. S. (2017). "The Mnemonic of Intuitive Ontology Violation Is Not the Distinctiveness Effect: Evidence from a Broad Age Spectrum of Persons in the UK and China during a Free-Recall Task." *Journal of Cognition and Culture, 17* (1–2), 169–97.

Greif, A. (1993). "Contract Enforceability and Economic Institutions in Early Trade: The Maghribi Traders' Coalition." *American Economic Review, 83* (3), 525–48.

Grice, H. P. (1991 [1967]). "Logic and Conversation" [1967, 1987]. In H. P. Grice (ed.), *Studies in the Way of Words*, pp. 1–143. Cambridge, Mass.: Harvard University Press.

Griskevicius, V., Tybur, J. M., Delton, A. W., and Robertson, T. E. (2011). "The Influence of Mortality and Socioeconomic Status on Risk and Delayed Rewards: A Life History Theory Approach." *Journal of Personality and Social Psychology, 100* (6), 1015–26. doi: 10.1037/a0022403

Griskevicius, V., Tybur, J. M., Gangestad, S. W., Shapiro, J. R., Kenrick, D. T., and Perea, E. F. (2009). "Aggress to Impress: Hostility as an Evolved Context-Dependent Strategy." *Journal of Personality and Social Psychology, 96* (5), 980–94.

Gueth, W., and van Damme, E. (1998). "Information, Strategic Behavior, and Fairness in Ultimatum Bargaining: An Experimental Study." *Journal of Mathematical Psychology, 42* (2–3), 227–47.

Gurven, M. (2004). "To Give and to Give Not: The Behavioral Ecology of Human Food Transfers." *Behavioral and Brain Sciences, 27* (4), 543–60.

Gurven, M., and Hill, K. (2009). "Why Do Men Hunt? A Reevaluation of 'Man the Hunter' and the Sexual Division of Labor." *Current Anthropology, 50* (1), 51–62.

Gurven, M., Hill, K., Kaplan, H. S., Hurtado, A., and Lyles, R. (2000). "Food Transfers among Hiwi Foragers of Venezuela: Tests of Reciprocity." *Human Ecology, 28* (2), 171–218.

Gurven, M., and Winking, J. (2008). "Collective Action in Action: Prosocial Behavior in and out of the Laboratory." *American Anthropologist, 110,* 179–90.

Guzman, R. A., and Munger, M. C. (2014). "Euvoluntariness and Just Market Exchange: Moral Dilemmas from Locke's Venditio." *Public Choice, 158* (1–2), 39–49. doi: http://link.springer.com/journal/volumesAndIssues/11127

Gwinner, E. (1996). "Circadian and Circannual Programmes in Avian Migration." *Journal of Experimental Biology, 199* (1), 39–48.

Hagen, E. H., and Hammerstein, P. (2006). "Game Theory and Human Evolution: A Critique of Some Recent Interpretations of Experimental Games." *Theoretical Population Biology* (3), 339–48.

Haidt, J. (2013). *The Righteous Mind: Why Good People Are Divided by Politics and Religion*. New York: Vintage Books.

Haidt, J., and Graham, J. (2007). "Planet of the Durkheimians, Where Community, Authority, and Sacredness Are Foundations of Morality." In *Social and Psychological Bases of Ideology and System Justification*, pp. 371–401. Oxford: Oxford University Press.

Haidt, J., and Joseph, C. (2004). "Intuitive Ethics: How Innately Prepared Intuitions Generate Culturally Variable Virtues." *Daedalus, 133* (4), 55–66.

Hamilton, W. D. (1963). "The Evolution of Altruistic Behavior." *American Naturalist, 97* (896), 354–56.

Hamlin, J. K., Wynn, K., and Bloom, P. (2007). "Social Evaluation by Preverbal Infants." *Nature, 450* (7169), 557–59.

Hanegraaff, W. J. (1998). *New Age Religion and Western Culture: Esotericism in the Mirror of Secular Thought*. Albany: State University of New York Press.

Hann, C. M., and Hart, K. (2011). *Economic Anthropology: History, Ethnography, Critique*. Cambridge: Polity Press.

Hannagan, R. J. (2008). "Gendered Political Behavior: A Darwinian Feminist Approach." *Sex Roles, 59* (7–8), 465–75. doi: 10.1007/s11199-008-9417-3

Harari, D., Gao, T., Kanwisher, N., Tenenbaum, J., and Ullman, S. (2016). "Measuring and Modeling the Perception of Natural and Unconstrained Gaze in Humans and Machines." arXiv preprint arXiv:1611.09819.

Hardin, R. (1982). *Collective Action*. Baltimore: Johns Hopkins University Press.

———. (1995). *One for All: The Logic of Group Conflict*. Princeton: Princeton University Press.

Harris, P. L. (1991). "The Work of the Imagination." In A. Whiten (ed.), *Natural Theories of Mind: Evolution, Development and Simulation of Everyday Mindreading*, pp. 283–304. Oxford: Blackwell.

Harris, P. L., and Lane, J. D. (2014). "Infants Understand How Testimony Works." *Topoi, 33* (2), 443–58. doi: 10.1007/s11245-013-9180-0

Haselton, M. G., and Gangestad, S. W. (2006). "Conditional Expression of Women's Desires and Men's Mate Guarding across the Ovulatory Cycle." *Hormones and Behavior, 49* (4), 509–18. doi: 10.1016/j.yhbeh.2005.10.006

Havel, V. (1985). "The Power of the Powerless." In J. Keane (ed.), *The Power of the Powerless: Citizens against the State in Central-Eastern Europe,* pp. 27–28. Armonk, N.Y.: Sharpe.

Hawkes, K. (2006). "Slow Life Histories and Human Evolution." In K. Hawkes and R. R. Paine (eds.), *The Evolution of Human Life History,* pp. 45–94. Santa Fe: School of American Research.

Hawkes, K., and Bliege Bird, R. (2002). "Showing Off, Handicap Signaling, and the Evolution of Men's Work." *Evolutionary Anthropology, 11,* 58–67.

Hechter, M. (1987). *Principles of Group Solidarity.* Berkeley: University of California Press.

Heine, B. (1997). *Possession: Cognitive Sources, Forces, and Grammaticalization.* Cambridge: Cambridge University Press.

Henrich, J., Fehr, E., et al. (2001). "In Search of Homo Economicus: Behavioral Experiments in 15 Small-Scale Societies." *American Economic Review, 91* (2), 73–78.

Herz, J. H. (2003). "The Security Dilemma in International Relations: Background and Present Problems." *International Relations 17,* (4), 411–16. doi: 10.1177/0047117803174001

Hibbing, J. R., Smith, K. B., and Alford, J. R. (2013). *Predisposed: Liberals, Conservatives, and the Biology of Political Difference.* London: Taylor and Francis.

Hilbig, B. E. (2009). "Sad, Thus True: Negativity Bias in Judgments of Truth." *Journal of Experimental Social Psychology, 45* (4), 983–86.

Hill, K., and Kaplan, H. S. (1999). "Life History Traits in Humans: Theory and Empirical Studies." *Annual Review of Anthropology, 28* (1), 397–430.

Hinde, R. A. (1987). *Individuals, Relationships and Culture: Links between Ethology and the Social Sciences.* Cambridge: Cambridge University Press.

Hirschfeld, L. A. (1994). "The Acquisition of Social Categories." In L. A. Hirschfeld and S. A. Gelman (eds.), *Mapping the Mind: Domain-Specificity in Culture and Cognition.* New York: Cambridge University Press.

———. (2013). "The Myth of Mentalizing and the Primacy of Folk Sociology." In M. R. Banaji and S. A. Gelman (eds.), *Navigating the Social World: What Infants, Children, and Other Species Can Teach Us,* pp. 101–6. New York: Oxford University Press.

Hirschfeld, L. A., and Gelman, S. A. (1994). *Mapping the Mind: Domain Specificity in Cognition and Culture.* Cambridge: Cambridge University Press.

Hitchner, R. B. (2005). "'The Advantages of Wealth and Luxury': The Case for Economic Growth in the Roman Empire." In I. Morris and J. Manning (eds.), *The Ancient Economy: Evidence and Models*, pp. 207–22. Stanford, Calif.: Stanford University Press.

Hobbes, T. (1651). *Leviathan, or, The matter, forme, and power of a common-wealth ecclesiasticall and civill*. London: Printed for Andrew Ckooke [i.e., Crooke], at the Green Dragon in St. Pauls Church-yard.

Hobsbawm, E. J., and Ranger, T. O. (1983). *The Invention of Tradition*. Cambridge: Cambridge University Press.

Hoffer, E. (1951). *The True Believer: Thoughts on the Nature of Mass Movements*. New York: Harper and Row.

Hombert, J.-M., and Ohala, J. J. (1982). "Historical Development of Tone Patterns." In J. P. Maher, A. R. Bomhard, and E. F. K. Koerner (eds.), *Papers from the 3rd International Conference on Historical Linguistics*, pp. 75–84. Amsterdam: Benjamins.

Hooker, C. I., Paller, K. A., Gitelman, D. R., Parrish, T. B., Mesulam, M. M., and Reber, P. J. (2003). "Brain Networks for Analyzing Eye Gaze." *Cognitive Brain Research, 17* (2), 406–18.

Hopkins, K. (1980). "Taxes and Trade in the Roman Empire (200 B.C.–A.D. 400)." *Journal of Roman Studies, 70*, 101–25.

Horne, C. (2001). "Sociologial Perspectives on the Emergence of Social Norms." In M. Hechter and K.-D. Opp (eds.), *Social Norms*, pp. 3–33. New York: Russell Sage Foundation.

Hornsey, M. J. (2008). "Social Identity Theory and Self-Categorization Theory: A Historical Review." *Social and Personality Psychology Compass, 2* (1), 204–22. doi: papers2://publication/doi/10.1111/j.1751-9004.2007.00066.x

Horowitz, D. L. (2001). *The Deadly Ethnic Riot*. Los Angeles: University of California Press.

Hrdy, S. (2009). *Mothers and Others: The Evolutionary Origins of Mutual Understanding*. Cambridge, Mass.: Belknap Press of Harvard University Press.

Hrdy, S. B. (1977). "Infanticide as a Primate Reproductive Strategy." *American Scientist, 65* (1), 40–49.

———. (1981). *The Woman That Never Evolved*. Cambridge, Mass.: Harvard University Press.

———. (2009). *Mothers and Others: The Evolutionary Origins of Mutual Understanding*. Cambridge, Mass.: Belknap Press of Harvard University Press.

Huang, P. C. (1996). *Civil Justice in China, Representation and Practice in the Qing*. Stanford, Calif.: Stanford University Press.

Human Rights Watch. (2008). *Perpetual Minors: Human Rights Abuses Stemming from Male Guardianship and Sex Segregation in Saudi Arabia.* New York: Human Rights Watch.

Humphrey, C., and Hugh-Jones, S. (1992). *Barter, Exchange and Value: An Anthropological Approach.* Cambridge: Cambridge University Press.

Hutchins, E. (1980). *Culture and Inference: A Trobriand Case Study.* Cambridge, Mass.: Harvard University Press.

Hyman, I. E., Jr., Husband, T. H., and Billings, F. J. (1995). "False Memories of Childhood Experiences." *Applied Cognitive Psychology, 9* (3), 181–97.

Ibn Khaldūn, M. (1958). *The Muqaddimah: An Introduction to History.* Trans. and ed. F. Rosenthal and N. J. Dawood. 3 vols. Princeton: Princeton University Press.

Irons, W. (2001). "Religion as a Hard-to-Fake Sign of Commitment." In R. Nesse (ed.), *Evolution and the Capacity for Commitment*, pp. 292–309. New York: Russell Sage Foundation.

Jaeggi, A. V., and Van Schaik, C. P. (2011). "The Evolution of Food Sharing in Primates." *Behavioral Ecology and Sociobiology, 65* (11), 2125–40. doi: 10.1007/s00265-011-1221-3

James, W. (1902). *The Varieties of Religious Experience: A Study in Human Nature: Being the Gifford Lectures on Natural Religion Delivered at Edinburgh in 1901–1902.* New York: Modern Library.

Jankowiak, W. R., and Fischer, E. (1992). "A Cross-Cultural Perspective on Romantic Love." *Ethnology* (31), 149–55.

Jaspers, K. (1953). *The Origin and Goal of History. (Vom Ursprung und Ziel der Geschichte, [1949].)* Trans. Michael Bullock. New Haven: Yale University Press.

Jayakody, R., and Kalil, A. (2002). "Social Fathering in Low-Income, African American Families with Preschool Children." *Journal of Marriage and Family, 64* (2), 504–16.

Jolly, M., and Thomas, N. (1992). "Introduction" [to special issue: *The Politics of Tradition in the Pacific*]. *Oceania, 62* (4), 241–48.

Jones, D. (2003). "The Generative Psychology of Kinship: Part 1: Cognitive Universals and Evolutionary Psychology." *Evolution and Human Behavior, 24* (5), 303–19.

Jordan, D. P. (1979). *The King's Trial: The French Revolution vs. Louis XVI.* Berkeley: University of California Press.

Jowett, A. (2014). "'But if you legalise same-sex marriage . . .': Arguments against Marriage Equality in the British Press." *Feminism and Psychology, 24* (1), 37–55. doi: 10.1177/0959353513510655

Jussim, L., Crawford, J. T., and Rubinstein, R. S. (2015). "Stereotype (In)accuracy in Perceptions of Groups and Individuals." *Current Directions in Psychological Science, 24* (6), 490–97. doi: 10.1177/0963721415605257

Jussim, L., Harber, K. D., Crawford, J. T., Cain, T. R., and Cohen, F. (2005). "Social Reality Makes the Social Mind: Self-Fulfilling Prophecy, Stereotypes, Bias, and Accuracy." *Interaction Studies: Social Behaviour and Communication in Biological and Artificial Systems, 6* (1), 85–102. doi: 10.1075/is.6.1.07jus

Kaiser, M. K., Jonides, J., and Alexander, J. (1986). "Intuitive Reasoning about Abstract and Familiar Physics Problems." *Memory and Cognition, 14,* 308–12.

Kalyvas, S. N. (2006). *The Logic of Violence in Civil War.* Cambridge: Cambridge University Press.

Kamble, S., Shackelford, T. K., Pham, M., and Buss, D. M. (2014). "Indian Mate Preferences: Continuity, Sex Differences, and Cultural Change across a Quarter of a Century." *Personality and Individual Differences, 70,* 150–55. doi: http://dx.doi.org/10.1016/j.paid.2014.06.024

Kant, I. (1781). *Critik der reinen Vernunft.* Riga: Johann Friedrich Hartknoch.

———. (1790). *Kritik der Urteilskraft.* Berlin: Libau.

Kaplan, H. S., and Gangestad, S. W. (2005). "Life History Theory and Evolutionary Psychology." In D. M. Buss (ed.), *The Handbook of Evolutionary Psychology.* Hoboken, N.J.: John Wiley and Sons.

Kaplan, H. S., and Gurven, M. (2005). "The Natural History of Human Food Sharing and Cooperation: A Review and a New Multi-Individual Approach to the Negotiation of Norms." In H. Gintis, S. Bowles, R. Boyd, and E. Fehr (eds.), *Moral Sentiments and Material Interests: The Foundations of Cooperation in Economic Life,* pp. 75–113. Cambridge, Mass.: MIT Press.

Kaplan, H. S., Hill, K., Lancaster, J., and Hurtado, A. (2000). "A Theory of Human Life History Evolution: Diet, Intelligence, and Longevity." *Evolutionary Anthropology, 9,* 156–85.

Karp, D., Jin, N., Yamagishi, T., and Shinotsuka, H. (1993). "Raising the Minimum in the Minimal Group Paradigm." *Japanese Journal of Experimental Social Psychology, 32* (3), 231–40.

Katz, P. R. (2009). *Divine Justice: Religion and the Development of Chinese Legal Culture.* Cambridge: Cambridge University Press.

Kaufmann, L., and Clément, F. (2007). "How Culture Comes to Mind: From Social Affordances to Cultural Analogies." *Intellectica, 46* (2–3), 221–50.

Keesing, R. (1984). "Rethinking 'Mana.'" *Journal of Anthropological Research, 40* (1), 137–56.

Keesing, R. M. (1993). "Kastom Re-examined." *Anthropological Forum, 6* (4), 587–96.

Kelemen, D. (2004). "Are Children 'Intuitive Theists'? Reasoning about Purpose and Design in Nature." *Psychological Science, 15* (5), 295–301.

Kelemen, D., and DiYanni, C. (2005). "Intuitions about Origins: Purpose and Intelligent Design in Children's Reasoning about Nature." *Journal of Cognition and Development, 6* (1), 3–31.

Kelemen, D., Seston, R., and Georges, L. S. (2012). "The Designing Mind: Children's Reasoning about Intended Function and Artifact Structure." *Journal of Cognition and Development, 13* (4), 439–53. doi: 10.1080/15248372.2011.608200

Kelly, R. L. (1995). *The Foraging Spectrum: Diversity in Hunter-Gatherer Lifeways.* Washington, D.C.: Smithsonian Institution Press.

Kenward, B., Karlsson, M., and Persson, J. (2011). "Over-Imitation Is Better Explained by Norm Learning Than by Distorted Causal Learning." *Proceedings of the Royal Society of London B: Biological Sciences, 278* (1709), 1239–46.

Kettell, S. (2013). "I Do, Thou Shalt Not: Religious Opposition to Same-Sex Marriage in Britain." *Political Quarterly, 84* (2), 247–55. doi: 10.1111/j.1467-923X.2013.12009.x

Keupp, S., Behne, T., Zachow, J., Kasbohm, A., and Rakoczy, H. (2015). "Over-Imitation Is Not Automatic: Context Sensitivity in Children's Over-Imitation and Action Interpretation of Causally Irrelevant Actions." *Journal of Experimental Child Psychology, 130,* 163–75.

KFF. (1996). "Survey of Americans and Economists on the Economy: The Washington Post/Kaiser Family Foundation/Harvard University Survey Project." Washington, D.C.: Kaiser Family Foundation.

Khan, T. S. (2006). *Beyond Honour: A Historical Materialist Explanation of Honour-Related Violence.* Oxford: Oxford University Press.

King, A. J., Johnson, D. D., and van Vugt, M. (2009). "The Origins and Evolution of Leadership." *Current Biology, 19,* R911–R916. doi: 10.1016/j.cub.2009.07.027

Kinzler, K. D., Shutts, K., Dejesus, J., and Spelke, E. S. (2009). "Accent Trumps Race in Guiding Children's Social Preferences." *Social Cognition, 27* (4), 623–34. doi: 10.1521/soco.2009.27.4.623

Kipnis, A. B. (1997). *Producing Guanxi: Sentiment, Self, and Subculture in a North China Village.* Durham, N.C.: Duke University Press.

Kirch, P. V. (2010). *How Chiefs Became Kings: Divine Kingship and the Rise of Archaic States in Ancient Hawai'i.* Berkeley: University of California Press.

Kiyonari, T., Tanida, S., and Yamagishi, T. (2000). "Social Exchange and Reciprocity: Confusion or a Heuristic?" *Evolution and Human Behavior, 21* (6), 411–27. doi: 10.1016/s1090-5138(00)00055-6

Klonoff, E. A., and Landrine, H. (1999). "Do Blacks Believe That HIV/AIDS Is a Government Conspiracy against Them?" *Preventive Medicine, 28* (5), 451–57. doi: http://dx.doi.org/10.1006/pmed.1999.0463

Komdeur, J. (2001). "Mate Guarding in the Seychelles Warbler Is Energetically Costly and Adjusted to Paternity Risk." *Proceedings of the Royal Society of London B: Biological Sciences, 268* (1481), 2103–11.

Kosmin, B. (2011). *One Nation under God: Religion in Contemporary America.* New York: Three Rivers Press.

Kramer, S. N. (1961). *Mythologies of the Ancient World* (1st ed.). Garden City, N.Y.: Doubleday.

Krasnow, M. M., Cosmides, L., Pedersen, E. J., and Tooby, J. (2012). "What Are Punishment and Reputation For?" *PLoS One, 7* (9), e45662.

Krasnow, M. M., Delton, A. W., Cosmides, L., and Tooby, J. (2016). "Looking under the Hood of Third-Party Punishment Reveals Design for Personal Benefit." *Psychological Science, 27* (3), 405–18.

Krasnow, M. M., Delton, A. W., Tooby, J., and Cosmides, L. (2013). "Meeting Now Suggests We Will Meet Again: Implications for Debates on the Evolution of Cooperation." *Nature Scientific Reports, 3,* 1747. doi: 10.1038/srep01747

Krebs, D., and Denton, K. (1997). "Social Illusions and Self-Deception: The Evolution of Biases in Person Perception." In J. A. Simpson, D. T. Kenrick, et al. (eds.), *Evolutionary Social Psychology,* pp. 21–48. Mahwah, N.J.: Lawrence Erlbaum.

Kropotkin, P. A. (1902). *Mutual Aid, a Factor of Evolution.* New York: McClure Phillips.

Kübler, S., Owenga, P., Reynolds, S. C., Rucina, S. M., and King, G. C. P. (2015). "Animal Movements in the Kenya Rift and Evidence for the Earliest Ambush Hunting by Hominins." *Scientific Reports, 5,* 14011. doi: 10.1038/srep14011

Kuhl, P. K., Stevens, E., Hayashi, A., Deguchi, T., Kiritani, S., and Iverson, P. (2006). "Infants Show a Facilitation Effect for Native Language Phonetic Perception between 6 and 12 Months." *Developmental Science, 9* (2), F13–F21. doi: 10.1111/j.1467-7687.2006.00468.x

Kuran, T. (1995). *Private Truths, Public Lies: The Social Consequences of Preference Falsification.* Cambridge, Mass.: Harvard University Press.

———. (1998). "Ethnic Norms and Their Transformation through Reputational Cascades." *Journal of Legal Studies, 27* (2), 623–59.

Kurzban, R., Descioli, P., and O'Brien, E. (2007). "Audience Effects on Moralistic Punishment." *Evolution and Human Behavior, 28* (2), 10.

Kurzban, R., McCabe, K., Smith, V. L., and Wilson, B. J. (2001). "Incremental Commitment and Reciprocity in a Real-Time Public Goods Game." *Personality and Social Psychology Bulletin, 27* (12), 1662–73.

Kurzban, R., and Neuberg, S. (2005). "Managing Ingroup and Outgroup Relationships." In D. M. Buss (ed.), *The Handbook of Evolutionary Psychology*, pp. 653–75. Hoboken, N.J.: John Wiley and Sons.

Kurzban, R., Tooby, J., and Cosmides, L. (2001). "Can Race Be Erased? Coalitional Computation and Social Categorization." *Proceedings of the National Academy of Sciences of the United States of America, 98* (26), 15387–92.

La Fontaine, J. S. (1998). *Speak of the Devil: Tales of Satanic Abuse in Contemporary England.* Cambridge: Cambridge University Press.

Labov, W. (1964). "Phonological Correlates of Social Stratification." *American Anthropologist, 66* (6, pt. 2), 164–76. doi: 10.1525/aa.1964.66.suppl_3.02a00120

Lakoff, G. (1987). *Women, Fire and Dangerous Things.* Chicago: University of Chicago Press.

Lakoff, G., and Johnson, M. (1980). *Metaphors We Live By.* Chicago: University of Chicago Press.

Lal, D. (2010). *Reviving the Invisible Hand: The Case for Classical Liberalism in the Twenty-First Century.* Princeton: Princeton University Press.

Landes, D. S. (1998). *The Wealth and Poverty of Nations: Why Some Are So Rich and Some So Poor* (1st ed.). New York: W. W. Norton.

Langlois, J. H., Kalakanis, L., Rubenstein, A. J., Larson, A., Hallam, M., and Smoot, M. (2000). "Maxims or Myths of Beauty? A Meta-Analytic and Theoretical Review." *Psychological Bulletin, 126* (3), 390–423. doi: 10.1037/0033-2909.126.3.390

Laurence, J., and Vaïsse, J. (2006). *Integrating Islam: Political and Religious Challenges in Contemporary France.* Washington, D.C.: Brookings Institution Press.

Le Roy Ladurie, E. (1975). "Famine Amenorrhoea (Seventeenth–Twentieth Centuries)." In R. Forster and O. Ranum (eds.), *Biology of Man in History*, pp. 163–78. Baltimore: Johns Hopkins University Press.

LeBlanc, S. A., and Register, K. E. (2003). *Constant Battles: The Myth of the Peaceful, Noble Savage* (1st ed.). New York: St. Martin's Press.

Lee, A. Y., Bond, G. D., Russell, D. C., Tost, J., González, C., and Scarbrough, P. S. (2010). "Team Perceived Trustworthiness in a Complex Military Peacekeeping Simulation." *Military Psychology, 22* (3), 237–61. doi: 10.1080/08995605.2010.492676

Lee, R. B. (1979). *The !Kung San: Men, Women, and Work in a Foraging Society.* Cambridge: Cambridge University Press.

Leeson, P. T., and Coyne, C. J. (2012). "Sassywood." *Journal of Comparative Economics, 40,* 608–20.

Leonardi, M., Gerbault, P., Thomas, M. G., and Burger, J. (2012). "The Evolution of Lactase Persistence in Europe: A Synthesis of Archaeological and Genetic

Evidence." *International Dairy Journal, 22* (2), 88–97. doi: http://dx.doi. org/10.1016/j.idairyj.2011.10.010

Leslie, A. M. (1987). "Pretense and Representation: The Origins of 'Theory of Mind.'" *Psychological Review, 94,* 412–26.

———. (1994). "ToMM, ToBy, and Agency: Core Architecture and Domain Specificity." In L. A. Hirschfeld and S. A. Gelman (eds.), *Mapping the Mind: Domain Specificity in Cognition and Culture,* pp. 119–48. New York: Cambridge University Press.

Leslie, A. M., Friedman, O., and German, T. P. (2004). "Core Mechanisms in 'Theory of Mind.'" *Trends in Cognitive Sciences, 8* (12), 529–33.

Lev-Ari, S., and Keysar, B. (2010). "Why Don't We Believe Non-Native Speakers? The Influence of Accent on Credibility." *Journal of Experimental Social Psychology.* doi: 10.1016/j.jesp.2010.05.025

Levine, N. E., and Silk, J. B. (1997). "Why Polyandry Fails: Sources of Instability in Polyandrous Marriages." *Current Anthropology, 38* (3), 375–98. doi: 10.1086/204624

Levitt, S., and List, J. (2007). "What Do Laboratory Experiments Measuring Social Preferences Reveal about the Real World?" *Journal of Economic Perspectives, 21* (2), 153–74.

Lewandowsky, S., Ecker, U. K. H., Seifert, C. M., Schwarz, N., and Cook, J. (2012). "Misinformation and Its Correction: Continued Influence and Successful Debiasing." *Psychological Science in the Public Interest, 13* (3), 106–31. doi: 10.1177/1529100612451018

Lewis, D. K. (1969). *Convention: A Philosophical Study.* Cambridge, Mass.: Harvard University Press.

Li, N. P., Bailey, J. M., Kenrick, D. T., and Linsenmeier, J. A. W. (2002). "The Necessities and Luxuries of Mate Preferences: Testing the Tradeoffs." *Journal of Personality and Social Psychology, 82* (6), 947–55. doi: 10.1037/0022-3514.82.6.947

Licht, L. E. (1976). "Sexual Selection in Toads (Bufo americanus)." *Canadian Journal of Zoology, 54* (8), 1277–84. doi: 10.1139/z76-145

Lieberman, D., Tooby, J., and Cosmides, L. (2007). "The Architecture of Human Kin Detection." *Nature, 445* (7129), 5.

Lieberman, M. D., Schreiber, D., and Ochsner, K. (2003). "Is Political Cognition Like Riding a Bicycle? How Cognitive Neuroscience Can Inform Research on Political Thinking." *Political Psychology, 24* (4), 681–704.

Lienard, P., Chevallier, C., Mascaro, O., Kiura, P., and Baumard, N. (2013). "Early Understanding of Merit in Turkana Children." *Journal of Cognition and Culture, 13* (1), 57–66.

Liesen, L. T. (2008). "The Evolution of Gendered Political Behavior: Contributions from Feminist Evolutionists." *Sex Roles, 59* (7–8), 476–81. doi: 10.1007/s11199-008-9465-8

List, J. A. (2007). "On the Interpretation of Giving in Dictator Games." *Journal of Political Economy, 115* (3), 482–93.

Lloyd, G. E. R. (2007). *Cognitive Variations: Reflections on the Unity and Diversity of the Human Mind.* New York: Oxford University Press.

Loftus, E. F. (1993). "The Reality of Repressed Memories." *American Psychologist, 48* (5), 518–37.

———. (1997). "Creating Childhood Memories." *Applied Cognitive Psychology, 11* (special issue), S75–S86.

———. (2005). "Planting Misinformation in the Human Mind: A 30-Year Investigation of the Malleability of Memory." *Learning and Memory, 12* (4), 361–66.

Low, B. S. (2000). *Why Sex Matters: A Darwinian Look at Human Behavior.* Princeton: Princeton University Press.

Luft, A. (2015). "Toward a Dynamic Theory of Action at the Micro Level of Genocide: Killing, Desistance, and Saving in 1994 Rwanda." *Sociological Theory, 33* (2), 148–72. doi: 10.1177/0735275115587721

Luhrmann, T. M. (2012). *When God Talks Back: Understanding the American Evangelical Relationship with God* (1st ed.). New York: Alfred A. Knopf.

Lumsden, C. J., and Wilson, E. O. (1981). *Genes, Minds and Culture.* Cambridge, Mass.: Harvard University Press.

Macfarlan, S. J., Walker, R. S., Flinn, M. V., and Chagnon, N. A. (2014). "Lethal Coalitionary Aggression and Long-Term Alliance Formation among Yanomamö Men." *Proceedings of the National Academy of Sciences, 111* (47), 16662–69. doi: 10.1073/pnas.1418639111

Mackay, C. (1841). *Memoirs of Extraordinary Popular Delusions and the Madness of Crowds.* London: Bentley.

Maeterlinck, M. (1930). *The Life of the Ant.* New York: John Day.

Major, B., Mendes, W. B., and Dovidio, J. F. (2013). "Intergroup Relations and Health Disparities: A Social Psychological Perspective." *Health Psychology, 32* (5), 514–24.

Malinowski, B. (1929). *The Sexual Life of Savages in Northwestern Melanesia: An Ethnographic Account of Courtship, Marriage and Family Life among the Natives of Trobriand Islands, British New Guinea.* New York: Harcourt, Brace.

Mallart Guimerà, L. (1981). *Ni dos ni ventre: Religion, magie et sorcellerie Evuzok.* Paris: Société d'ethnographie.

———. (2003). *La forêt de nos ancêtres.* Tervuren, Belgium: Musée royal de l'Afrique centrale.

Mann, S. E. (1955). *Ancient Near Eastern Texts Relating to the Old Testament.* Princeton: Princeton University Press.

Marcus, G. E. (2013). *Political Psychology: Neuroscience, Genetics and Politics.* New York: Oxford University Press.

Markson, L., and Bloom, P. (1997). "Evidence against a Dedicated System for Word Learning in Children." *Nature, 385* (6619), 813–15. doi: 10.1038/385813a0

Marlowe, F. (2000). "Paternal Investment and the Human Mating System." *Behavioural Processes, 51* (1–3), 45–61.

Martin, L. H. (1987). *Hellenistic Religions: An Introduction.* Oxford: Oxford University Press.

Maryanski, A., and Turner, J. H. (1992). *The Social Cage: Human Nature and the Evolution of Society.* Stanford, Calif.: Stanford University Press.

Mascaro, O., and Sperber, D. (2009). "The Moral, Epistemic, and Mindreading Components of Children's Vigilance Towards Deception." *Cognition, 112* (3), 367–80.

Mather, C. (2005). "Accusations of Genital Theft: A Case from Northern Ghana." *Culture, Medicine and Psychiatry, 29* (1), 33–52. doi: 10.1007/s11013-005-4622-9

Mauss, M. (1973 [1937]). "Techniques of the Body." *Economy and Society, 2,* 70–88.

Maynard Smith, J. (1964). "Group Selection and Kin Selection." *Nature, 201* (4924), 1145–47.

———. (1982). *Evolution and the Theory of Games.* Cambridge: Cambridge University Press.

Maynard Smith, J., and Harper, D. (2003). *Animal Signals* (1st ed.). New York: Oxford University Press.

McCabe, K. A., and Smith, V. L. (2001). "Goodwill Accounting and the Process of Exchange." In G. Gigerenzer and R. Selten (eds.), *Bounded Rationality: The Adaptive Toolbox,* pp. 319–40. Cambridge, Mass.: MIT Press.

McCaffree, K. (2017). *The Secular Landscape: The Decline of Religion in America.* New York: Palgrave Macmillan.

McCauley, R. N. (2011). *Why Religion Is Natural and Science Is Not.* Oxford: Oxford University Press.

McCauley, R. N., and Lawson, E. T. (1984). "Functionalism Reconsidered." *History of Religions, 23,* 372–81.

———. (2002). *Bringing Ritual to Mind: Psychological Foundations of Cultural Forms.* Cambridge: Cambridge University Press.

McCloskey, D. N. (2006). *The Bourgeois Virtues: Ethics for an Age of Commerce.* Chicago: University of Chicago Press.

McDermott, R. (2004). *Political Psychology in International Relations.* Ann Arbor: University of Michigan Press.

———. (2011). "Hormones and Politics." In P. K. Hatemi and R. McDermott (eds.), *Man Is by Nature a Political Animal,* pp. 247–60. Chicago: University of Chicago Press.

McDonald, M., Navarrete, C. D., and van Vugt, M. (2012). "Evolution and the Psychology of Intergroup Conflict: The Male Warrior Hypothesis." *Philosophical Transactions of the Royal Society B, 367,* 670–79. doi: 10.1098/rstb.2011.0301

McGarty, C., Yzerbyt, V. Y., and Spears, R. (2002). *Stereotypes as Explanations: The Formation of Meaningful Beliefs about Social Groups.* New York: Cambridge University Press.

Medina, L. F. (2007). *A Unified Theory of Collective Action and Social Change.* Ann Arbor: University of Michigan Press.

Mendes, W. B., Blascovich, J., Lickel, B., and Hunter, S. (2002). "Challenge and Threat during Social Interaction with White and Black Men." *Personality and Social Psychology Bulletin, 28*(7), 939–52. doi: 10.1177/01467202028007007

Mendle, J., Harden, K. P., Turkheimer, E., Van Hulle, C. A., D'Onofrio, B. M., Brooks-Gunn, J., . . . Lahey, B. B. (2009). "Associations between Father Absence and Age of First Sexual Intercourse." *Child Development, 80* (5), 1463–80. doi: 10.1111/j.1467-8624.2009.01345.x

Mercier, H. (2017). "How Gullible Are We? A Review of the Evidence from Psychology and Social Science." *Review of General Psychology, 21* (2), 103. doi: dx.doi.org/10.1037/gpr0000111

Mercier, H., and Sperber, D. (2011). "Why Do Humans Reason? Arguments for an Argumentative Theory." *Behavioral and Brain Sciences, 34* (2), 57–74. doi: 10.1017/s0140525x10000968

———. (2017). *The Enigma of Reason.* Cambridge, Mass.: Harvard University Press.

Mernissi, F. (1987). *Beyond the Veil: Male-Female Dynamics in Modern Muslim Society.* Bloomington: Indiana University Press.

Merton, R. K. (1996). *On Social Structure and Science.* Chicago: University of Chicago Press.

Mesnick, S. L. (1997). "Sexual Alliances: Evidence and Evolutionary Implications." In P. Gowaty (ed.), *Feminism and Evolutionary Biology,* pp. 207–60. Dordrecht: Springer.

Miklósi, A., Polgárdi, R., Topál, J., and Csányi, V. (1998). "Use of Experimenter-Given Cues in Dogs." *Animal Cognition, 1,* 113–22.

Milinski, M., Semmann, D., and Krambeck, H. J. (2002). "Reputation Helps Solve the 'Tragedy of the Commons.'" *Nature, 415* (6870), 424–26.

Miller, G. F., and Todd, P. M. (1998). "Mate Choice Turns Cognitive." *Trends in Cognitive Sciences, 2* (5), 190–98. doi: 10.1016/S1364-6613(98)01169-3

Miner, E. J., Shackelford, T. K., and Starratt, V. G. (2009). "Mate Value of Romantic Partners Predicts Men's Partner-Directed Verbal Insults." *Personality and Individual Differences, 46* (2), 135–39. doi: http://dx.doi.org/10.1016/j.paid.2008.09.015

Mitchell, R. (1986). "A Framework for Discussing Deception." In R. Mitchell and N. Thompson (eds.), *Deception: Perspectives on Human and Nonhuman Deceit,* pp. 3–40. Albany: SUNY Press.

Mokyr, J. (1992). *The Lever of Riches: Technological Creativity and Economic Progress.* Oxford: Oxford University Press.

Moon, C., Lagercrantz, H., and Kuhl, P. K. (2013). "Language Experienced in Utero Affects Vowel Perception after Birth: A Two-Country Study." *Acta Paediatrica, 102* (2), 156–60. doi: 10.1111/apa.12098

Morin, O. (2016). *How Traditions Live and Die.* Oxford: Oxford University Press.

Morris, I. (2006). "The Growth of Greek Cities in the First Millennium B C ." In G. Storwey (ed.), *Urbanism in the Preindustrial World: Cross-Cultural Approaches,* pp. 26–51. Tuscaloosa: University of Alabama Press.

———. (2013). *The Measure of Civilization: How Social Development Decides the Fate of Nations.* Princeton: Princeton University Press.

Mueller, J. E. (2004). *The Remnants of War.* Ithaca: Cornell University Press.

Munger, M. C. (2010). "Endless Forms Most Beautiful and Most Wonderful: Elinor Ostrom and the Diversity of Institutions." *Public Choice, 143* (3–4), 263–68. doi: http://link.springer.com/journal/volumesAndIssues/11127

———. (2015). *Choosing in Groups: Analytical Politics Revisited.* Cambridge: Cambridge University Press.

Musolino, J. (2015). *The Soul Fallacy: What Science Shows We Gain from Letting Go of Our Soul Beliefs.* New York: Prometheus Books.

Nagell, K., Olguin, R. S., and Tomasello, M. (1993). "Processes of Social Learning in the Tool Use of Chimpanzees (Pan troglodytes) and Human Children (Homo sapiens)." *Journal of Comparative Psychology, 107* (2), 174.

Needham, R. (1971). "Remarks on the Analysis of Kinship and Marriage." In R. Needham (ed.), *Rethinking Kinship and Marriage,* pp. 1–33. London: Tavistock.

———. (1972). *Belief, Language, and Experience.* Chicago: University of Chicago Press.

Nell, V. (2006). "Cruelty's Rewards: The Gratification of Perpetrators and Spectators." *Behavioral and Brain Sciences, 29,* 211–57.

Nereid, C. T. (2011). "Kemalism on the Catwalk: The Turkish Hat Law of 1925." *Journal of Social History, 44* (3), 707–28.

Nesdale, D., and Rooney, R. (1996). "Evaluations and Stereotyping of Accented Speakers by Pre-Adolescent Children." *Journal of Language and Social Psychology, 15* (2), 133–54. doi: 10.1177/0261927x960152002

Nettle, D. (2010). "Dying Young and Living Fast: Variation in Life History across English Neighborhoods." *Behavioral Ecology, 21* (2), 387–95.

Nettle, D., Coall, D. A., and Dickins, T. E. (2011). "Early-Life Conditions and Age at First Pregnancy in British Women." *Proceedings: Biological Sciences / The Royal Society, 278* (1712), 1721–27. doi: 10.1098/rspb.2010.1726

Nettle, D., Colléony, A., and Cockerill, M. (2011). "Variation in Cooperative Behaviour within a Single City." *PLoS One, 6* (10), e26922.

Nettle, D., Grace, J. B., Choisy, M., Cornell, H. V., Guégan, J. F., and Hochberg, M. E. (2007). "Cultural Diversity, Economic Development and Societal Instability." *PLoS One, 2* (9), e929.

Neuberg, S. L., Kenrick, D. T., and Schaller, M. (2010). "Evolutionary Social Psychology." In S. T. Fiske, D. T. Gilbert, and G. Lindzey (eds.), *Handbook of Social Psychology,* vol. 2 (5th ed.), pp. 761–96. Hoboken, N.J.: John Wiley and Sons.

———. (2011). "Human Threat Management Systems: Self-Protection and Disease Avoidance." *Neuroscience and Biobehavioral Reviews, 35* (4), 1042–51. doi: 10.1016/j.neubiorev.2010.08.011

Nietzsche, F. (1882). *Die fröhliche Wissenschaft.* Lepizig: Fritzsch.

———. (1980 [1901]). *Der Wille zur Macht: Versuch einer Umwertung aller Werte* (12th ed.). Stuttgart: Kröner.

Noë, R., and Hammerstein, P. (1994). "Biological Markets: Supply and Demand Determine the Effect of Partner Choice in Cooperation, Mutualism and Mating." *Behavioral Ecology and Sociobiology, 35* (1), 1–11. doi: 10.1007/bf00167053

Noë, R., van Schaik, C., and Van Hooff, J. (1991). "The Market Effect: An Explanation for Pay-Off Asymmetries among Collaborating Animals." *Ethology, 87* (1–2), 97–118.

Noles, N. S., and Keil, F. (2011). "Exploring Ownership in a Developmental Context." In H. H. Ross and O. Friedman (eds.), *Origins of Ownership of Property: New Directions for Child and Adolescent Development,* vol. 132, pp. 91–103. New York: Wiley.

Norenzayan, A., and Shariff, A. F. (2008). "The Origin and Evolution of Religious Prosociality." *Science, 322,* 58–61.

Norwich, J. J. (1989). *Byzantium: The Early Centuries.* New York: Knopf.

Nozick, R. (1974). *Anarchy, State, and Utopia*. New York: Basic Books.

Offit, P. A. (2011). *Deadly Choices: How the Anti-Vaccine Movement Threatens Us All*. New York: Basic Books.

Öhman, A., Flykt, A., and Esteves, F. (2001). "Emotion Drives Attention: Detecting the Snake in the Grass." *Journal of Experimental Psychology: General, 130* (3), 466–78.

Öhman, A., and Mineka, S. (2001). "Fears, Phobias, and Preparedness: Toward an Evolved Module of Fear and Fear Learning." *Psychological Review, 108* (3), 483–522.

Olson, M. (1965). *The Logic of Collective Action: Public Goods and the Theory of Groups*. Cambridge, Mass.: Harvard University Press.

Onishi, K. H., and Baillargeon, R. (2005). "Do 15-Month-Old Infants Understand False Beliefs?" *Science, 308* (5719), 255.

Ostrom, E. (1990). *Governing the Commons: The Evolution of Institutions for Collective Action*. Cambridge: Cambridge University Press.

———. (2005). *Understanding Institutional Diversity*. Princeton: Princeton University Press.

Otto, R. (1920). *Das Heilige* (4th ed.). Breslau: Trewendt und Granier.

Oxley, D. R., Smith, K. B., Alford, J. R., Hibbing, M. V., Miller, J. L., Scalora, M., . . . Hibbing, J. R. (2008). "Political Attitudes Vary with Physiological Traits." *Science, 321* (5896), 1667–70.

Padoa-Schioppa, C., and Assad, J. A. (2006). "Neurons in the Orbitofrontal Cortex Encode Economic Value." *Nature, 441* (7090), 223–26. doi: http://www.nature.com/nature/journal/v441/n7090/suppinfo/nature04676_S1.html

Page-Gould, E., Mendoza-Denton, R., and Tropp, L. R. (2008). "With a Little Help from My Cross-Group Friend: Reducing Anxiety in Intergroup Contexts through Cross-Group Friendship." *Journal of Personality and Social Psychology, 95* (5), 1080–94.

Paladino, M.-P., and Castelli, L. (2008). "On the Immediate Consequences of Intergroup Categorization: Activation of Approach and Avoidance Motor Behavior toward Ingroup and Outgroup Members." *Personality and Social Psychology Bulletin, 34* (6), 755–68. doi: 10.1177/0146167208315155

Payne, B. K. (2001). "Prejudice and Perception: The Role of Automatic and Controlled Processes in Misperceiving a Weapon." *Journal of Personality and Social Psychology, 81* (2), 181–92. doi: 10.1037/0022-3514.81.2.181

Payne, B. K., Lambert, A. J., and Jacoby, L. L. (2002). "Best Laid Plans: Effects of Goals on Accessibility Bias and Cognitive Control in Race-Based Misperceptions of Weapons." *Journal of Experimental Social Psychology, 38* (4), 384–96. doi: 10.1016/s0022-1031(02)00006-9

Pelphrey, K. A., Morris, J. P., and McCarthy, G. (2005). "Neural Basis of Eye Gaze Processing Deficits in Autism." *Brain: A Journal of Neurology, 128* (5), 1038–48.

Perrett, D. I., Burt, D. M., Penton-Voak, I. S., Lee, K. J., Rowland, D. A., and Edwards, R. (1999). "Symmetry and Human Facial Attractiveness." *Evolution and Human Behavior, 20* (5), 295–307.

Peters, E. L. (1978). "The Status of Women in Four Middle East Communities." In L. Beck and N. Kiddie (eds.), *Women in the Muslim World,* pp. 311–50. Cambridge, Mass.: Harvard University Press.

Pettigrew, T. F., and Tropp, L. R. (2008). "How Does Intergroup Contact Reduce Prejudice? Meta-Analytic Tests of Three Mediators." *European Journal of Social Psychology, 38* (6), 922–34.

Pettit, P. (2003). "Groups with Minds of Their Own." In F. Schmitt (ed.), *Socializing Metaphysics: The Nature of Social Reality,* pp. 167–93. Lanham, Md.: Rowman and Littlefield.

Pew Research Center. (2013). "The World's Muslims: Religion, Politics and Society." *Pew Research Center's Forum on Religion and Public Life,* pp. 226–37. Washington, D.C.: Pew Research Center.

Piaget, J. (1932). *The Moral Judgment of the Child.* London: Routledge and Kegan Paul.

Piazza, J., and Bering, J. (2008). "Concerns about Reputation Via Gossip Promote Generous Allocations in an Economic Game." *Evolution and Human Behavior, 29* (3), 172–78.

Pickrell, J. K., Coop, G., Novembre, J., Kudaravalli, S., Li, J. Z., Absher, D., . . . Pritchard, J. K. (2009). "Signals of Recent Positive Selection in a Worldwide Sample of Human Populations." *Genome Research, 19* (5), 826–37. doi: 10.1101/gr.087577.108

Pietraszewski, D. (2013). "What Is Group Psychology? Adaptations for Mapping Shared Intentional Stances." In M. R. Banaji, S. A. Gelman, M. R. Banaji, and S. A. Gelman (eds.), *Navigating the Social World: What Infants, Children, and Other Species Can Teach Us,* pp. 253–57. New York: Oxford University Press.

Pietraszewski, D., Cosmides, L., and Tooby, J. (2014). "The Content of Our Cooperation, Not the Color of Our Skin: An Alliance Detection System Regulates Categorization by Coalition and Race, but Not Sex." *PLoS One, 9* (2), 1–19. doi: 10.1371/journal.pone.0088534

Pietraszewski, D., Curry, O. S., Petersen, M. B., Cosmides, L., and Tooby, J. (2015). "Constituents of Political Cognition: Race, Party Politics, and the Alliance Detection System." *Cognition, 140,* 24–39. doi: 10.1016/j.cognition.2015.03.007

Pike, S. M. (2012). *New Age and Neopagan Religions in America*. New York: Columbia University Press.

Pinker, S. (1984). *Language Learnability and Language Development*. Cambridge, Mass.: Harvard University Press.

———. (1989). *Learnability and Cognition: The Acquisition of Argument Structure*. Cambridge, Mass.: MIT Press.

Plott, C. R. (1974). "On Game Solutions and Revealed Preference Theory." *Social Science Working Papers, California Institute of Technology*. (Vol. 35). Pasadena: California Institute of Technology.

Polanyi, K. (2001[1957]). "The Economy as Instituted Process." In M. Granovetter and R. Swedberg (eds.), *The Sociology of Economic Life*. Boulder, Colo.: Westview Press.

Porta, D. (2008). "Research on Social Movements and Political Violence." *Qualitative Sociology, 31* (3), 221–30.

Posner, E. A. (2000). *Law and Social Norms*. Cambridge, Mass.: Harvard University Press.

Posner, R. A. (1980). "A Theory of Primitive Society, with Special Reference to Law." *Journal of Law and Economics, 23* (1), 1–53.

———. (2001). "A Theory of Primitive Society, with Special Reference to Law." In F. Parisi (ed.), *The Collected Essays of Richard A. Posner*, vol. 2: *The Economics of Private Law*, pp. 3–55. Northampton, Mass.: Elgar.

Povinelli, D. J. (2003). "Folk Physics for Apes: The Chimpanzee's Theory of How the World Works." *Human Development, 46*, 161–68.

Povinelli, D. J., and Eddy, T. J. (1996). "Chimpanzees: Joint Visual Attention." *Psychological Science, 7* (3), 129–35.

Pratto, F., and John, O. P. (1991). "Automatic Vigilance: The Attention-Grabbing Power of Negative Social Information." *Journal of Personality and Social Psychology, 61* (3), 380–91.

Price, M. E. (2005). "Punitive Sentiment among the Shuar and in Industrialized Societies: Cross-Cultural Similarities." *Evolution and Human Behavior, 26* (3), 279–87. doi: 10.1016/j.evolhumbehav.2004.08.009

Putnam, R. D. (2000). *Bowling Alone: The Collapse and Revival of American Community*. New York: Simon and Schuster.

———. (2007). "E Pluribus Unum: Diversity and Community in the Twenty-First Century." The 2006 Johan Skytte Prize Lecture. *Scandinavian Political Studies, 30* (2), 137–74.

——— (ed.). (2002). *Democracies in Flux: The Evolution of Social Capital in Contemporary Society*. New York: Oxford University Press.

———. (2010). *American Grace: How Religion Divides and Unites Us*. New York: Simon and Schuster.

Puts, D. A. (2005). "Mating Context and Menstrual Phase Affect Women's Preference for Male Voice Pitch." *Evolution and Human Behavior, 31,* 157–75.

Quigley, D. (1993). *The Interpretation of Caste.* Oxford: Clarendon Press.

———. (2005). *The Character of Kingship* (U.K. ed.). Oxford: Berg.

Quinlan, R. J. (2003). "Father Absence, Parental Care, and Female Reproductive Development." *Evolution and Human Behavior, 24* (6), 376–90.

Quinlan, R. J., Quinlan, M. B., and Flinn, M. V. (2003). "Parental Investment and Age at Weaning in a Caribbean Village." *Evolution and Human Behavior, 24* (1), 1–16.

Rabbie, J., Schot, J., and Visser, L. (1989). "Social Identity Theory: A Conceptual Critique from the Perspective of a Behavioural Interaction Model." *European Journal of Social Psychology, 19,* 171–202.

Rachman, S. (1977). "The Conditioning Theory of Fear-Acquisition: A Critical Examination." *Behaviour Research and Therapy, 15* (5), 375–87. doi: 10.1016/0005-7967(77)90041-9

Rakoczy, H., and Schmidt, M. F. H. (2013). "The Early Ontogeny of Social Norms." *Child Development Perspectives, 7* (1), 17–21. doi: 10.1111/cdep.12010

Rakoczy, H., Warneken, F., and Tomasello, M. (2008). "The Sources of Normativity: Young Children's Awareness of the Normative Structure of Games." *Developmental Psychology, 44* (3), 875–81. doi: 10.1037/0012-1649.44.3.87510 .1037/0012-1649.44.3.875.supp (Supplemental).

Ramble, C. (2008). *The Navel of the Demoness: Tibetan Buddhism and Civil Religion in Highland Nepal.* Oxford: Oxford University Press.

Rawls, J. (1971). *A Theory of Justice.* Cambridge, Mass.: Belknap Press of Harvard University Press.

Read, L. E., Reed, L. W., Ebeling, R. M., and Friedman, M. (2009). *I, Pencil.* Atlanta: Foundation for Economic Education.

Renfrew, C. (1969). "Trade and Culture Process in European Prehistory." *Current Anthropology, 10* (2–3), 151–69.

Rhodes, G., Proffitt, F., Grady, J. M., and Sumich, A. (1998). "Facial Symmetry and the Perception of Beauty." *Psychonomic Bulletin and Review, 5* (4), 659–69.

Ricardo, D. (1817). *On the Principles of Political Economy, and Taxation.* London: John Murray at John M'Creery's Printers.

Richerson, P. J., and Boyd, R. (2005). *Not by Genes Alone: How Culture Transformed Human Evolution.* Chicago: University of Chicago Press.

Ridley, M. (1996). *The Origins of Virtue: Human Instincts and the Evolution of Cooperation.* New York: Penguin Books.

———. (2010). *The Rational Optimist: How Prosperity Evolves* (1st U.S. ed.). New York: Harper.

Roemer, J. E. (1996). *Theories of Distributive Justice*. Cambridge, Mass.: Harvard University Press.

Roff, D. A. (2007). "Contributions of Genomics to Life-History Theory." *Nature Reviews Genetics, 8* (2), 116–25.

Rosenberg, A. (1980). *Sociobiology and the Preemption of Social Science*. Baltimore: Johns Hopkins University Press.

Rosenblum, L. A., and Paully, G. S. (1984). "The Effects of Varying Environmental Demands on Maternal and Infant Behavior." *Child Development, 55* (1), 305. doi: 10.1111/1467-8624.ep7405592

Ross, L., Greene, D., and House, P. (1977). "The False Consensus Effect: An Egocentric Bias in Social Perception and Attribution Processes." *Journal of Experimental Social Psychology, 13* (3), 279–301.

Rotberg, R. (1999). "Social Capital and Political Culture in Africa, America, Australasia, and Europe." *Journal of Interdisciplinary History, 29* (3), 339–56.

Roth, I. (2007). *Imaginative Minds*. Oxford: New York.

Rothbart, M., and Taylor, M. (1990). "Category Labels and Social Reality: Do We View Social Categories as Natural Kinds?" In K. F. G. Semin (ed.), *Language and Social Cognition*, pp. 11–36. London: Sage.

Rothschild, L. S. (2001). *Psychological Essentialism: Social Categories, and the Impact of Prejudice*. New York: New School for Social Research.

Rousseau, J.-J. (1984 [1755]). *A Discourse on Inequality*. Trans. M. Cranston. Harmondsworth, U.K.: Penguin.

———. (1762). *Du contrat social*. Chicoutimi, Québec: Les classiques en sciences sociales.

Rowe, D. C. (2002). "On Genetic Variation in Menarche and Age at First Sexual Intercourse: A Critique of the Belsky-Draper Hypothesis." *Evolution and Human Behavior, 23* (5), 365–72. doi: 10.1016/S1090-5138(02)00102-2

Rozin, P., Millman, L., and Nemeroff, C. (1986). "Operation of the Laws of Sympathetic Magic in Disgust and Other Domains." *Journal of Personality and Social Psychology, 50* (4), 703–12.

Rozin, P., and Royzman, E. B. (2001). "Negativity Bias, Negativity Dominance, and Contagion." *Personality and Social Psychology Review, 5* (4), 296–320. doi: 10.1207/S15327957PSPR0504_2

Rubin, P. H. (2002). *Darwinian Politics: The Evolutionary Origin of Freedom*. New Brunswick, N.J.: Rutgers University Press.

———. (2013). *Emporiophobia (Fear of Markets): Cooperation or Competition?* New York: Technology Policy Institute.

Ryan, M. J., and Guerra, M. A. (2014). "The Mechanism of Sound Production in Túngara Frogs and Its Role in Sexual Selection and Speciation." *Current*

Opinion in Neurobiology, 28, 54–59. doi: http://dx.doi.org/10.1016/j.conb.2014.06.008

Saad, G. (2012). *The Evolutionary Bases of Consumption.* Mahwah, N.J.: Lawrence Erlbaum.

Saler, B., Ziegler, C. A., and Moore, C. B. (1997). *UFO Crash at Roswell: The Genesis of a Modern Myth.* Washington, D.C.: Smithsonian Institution.

Sanderson, S. K. (2014). *Human Nature and the Evolution of Society.* Boulder, Colo.: Westview Press.

Santos, L. R., and Platt, M. L. (2014). "Evolutionary Anthropological Insights into Neuroeconomics: What Non-Human Primates Can Tell Us about Human Decision-Making Strategies." In E. Fehr and P. W. Glimcher (eds.), *Neuroeconomics* (2nd ed.), pp. 109–22. San Diego: Academic Press.

Sarkissian, H., Chatterjee, A., De Brigard, F., Knobe, J., Nichols, S., and Sirker, S. (2010). "Is Belief in Free Will a Cultural Universal?" *Mind and Language, 25* (3), 346–58. doi: 10.1111/j.1468-0017.2010.01393.x

Scheidel, W., and Friesen, S. J. (2009). "The Size of the Economy and the Distribution of Income in the Roman Empire." *Journal of Roman Studies, 99,* 61–91.

Schelling, T. C. (1971). "Dynamic Models of Segregation." *Journal of Mathematical Sociology, 1,* 143–86.

———. (1978). *Micromotives and Macrobehavior* (1st ed.). New York: W. W. Norton.

Schmitt, D. P. (2003). "Universal Sex Differences in the Desire for Sexual Variety: Tests from 52 Nations, 6 Continents, and 13 Islands." *Journal of Personality and Social Psychology, 85* (1), 85–104. doi: 10.1037/0022-3514.85.1.85

Scott-Phillips, T. C. (2008). "Defining Biological Communication." *Journal of Evolutionary Biology, 21* (2), 387–95. doi: 10.1111/j.1420-9101.2007.01497.x

Scribner, R. (1990). "Politics and the Institutionalisation of Reform in Germany." In G. Elton (ed.), *The New Cambridge Modern History,* vol. 2: *The Reformation,* pp. 172–97. Cambridge: Cambridge University Press.

Seabright, P. (2010). *The Company of Strangers: A Natural History of Economic Life.* Princeton: Princeton University Press.

———. (2012). *The War of the Sexes: How Conflict and Cooperation Have Shaped Men and Women from Prehistory to the Present.* Princeton: Princeton University Press.

Searcy, W. A., and Nowicki, S. (2010). *The Evolution of Animal Communication: Reliability and Deception in Signaling Systems.* Princeton: Princeton University Press.

Sears, D. O., Huddy, L., and Jervis, R. (2003). *Oxford Handbook of Political Psychology.* New York: Oxford University Press.

Sell, A. (2011). "The Recalibrational Theory and Violent Anger." *Aggression and Violent Behavior, 16* (5), 381–89. doi: 10.1016/j.avb.2011.04.013

Sell, A., Tooby, J., and Cosmides, L. (2009). "Formidability and the Logic of Human Anger." *PNAS Proceedings of the National Academy of Sciences of the United States of America, 106* (35), 15073–78. doi: 10.1073/pnas.0904312106

Sellars, W. (1963 [1991]). *Science, Perception and Reality.* London: Routledge and Kegan Paul and Humanities Press.

Sen, A. K. (2009). *The Idea of Social Justice.* Cambridge, Mass.: Belknap Press of Harvard University Press.

Service, E. R. (1965). *Primitive Social Organization: An Evolutionary Perspective.* New York: Random House.

Setchell, J. M., Charpentier, M., and Wickings, E. J. (2005). "Mate Guarding and Paternity in Mandrills: Factors Influencing Alpha Male Monopoly." *Animal Behaviour, 70* (5), 1105–20. doi: http://dx.doi.org/10.1016/j.anbehav.2005.02.021

Seyfarth, R. M., and Cheney, D. L. (2003). "Signalers and Receivers in Animal Communication." *Annual Review of Psychology, 54,* 145–73.

Sharer, R. J., and Traxler, L. P. (2006). *The Ancient Maya.* Stanford, Calif.: Stanford University Press.

Sharf, R. H. (1998). "Experience." In M. C. Taylor (ed.), *Critical Terms in Religious Studies,* pp. 94–116. Chicago: University of Chicago Press.

———. (2000). "The Rhetoric of Experience and the Study of Religion." *Journal of Consciousness Studies, 7* (11–12), 267–87.

Shariff, A. F., and Norenzayan, A. (2011). "Mean Gods Make Good People: Different Views of God Predict Cheating Behavior." *International Journal for the Psychology of Religion, 21* (2), 85–96. doi: 10.1080/10508619.2011.556990

Sheehy, P. (2012). *The Reality of Social Groups.* London: Routledge.

Sidanius, J., and Pratto, F. (1999). *Social Dominance: An Intergroup Theory of Social Oppression and Hierarchy.* Cambridge: Cambridge University Press.

Sidanius, J., and Veniegas, R. C. (2000). "Gender and Race Discrimination: The Interactive Nature of Disadvantage." In S. Oskamp et al. (eds.), *Reducing Prejudice and Discrimination,* pp. 47–69. Mahwah, N.J.: Lawrence Erlbaum.

Slingerland, E. G. (2007). *Effortless Action: Wu-Wei as Conceptual Metaphor and Spiritual Ideal in Early China.* New York: Oxford University Press.

———. (2008). *What Science Offers the Humanities: Integrating Body and Culture.* New York: Cambridge University Press.

Slingerland, E. G., and Collard, M. (2012). *Creating Consilience: Integrating the Sciences and the Humanities.* New York: Oxford University Press.

Slone, D. J. (2004). *Theological Incorrectness: Why Religious People Believe What They Shouldn't.* Oxford: Oxford University Press.

Smith, A. (1767). *The theory of moral sentiments. To which is added a dissertation on the origin of languages* (3d ed.). London: Printed for A. Millar, A. Kincaid and J. Bell in Edinburgh; and sold by T. Cadell.

———. (1776). *An inquiry into the nature and causes of the wealth of nations.* London: Printed for W Strahan and T Cadell, in the Strand, London.

Smith, A. D. (1987). *The Ethnic Origins of Nations.* Oxford: Basil Blackwell.

Smith, A. G. (2008). "The Implicit Motives of Terrorist Groups: How the Needs for Affiliation and Power Translate into Death and Destruction." *Political Psychology, 29* (1), 55–75.

Smith, E. A. (1998). "Is Tibetan Polyandry Adaptive? Methodological and Metatheoretical Analyses." *Human Nature, 9* (3), 225–61.

Smith, R. M. (2003). *Stories of Peoplehood: The Politics and Morals of Political Membership.* Cambridge: Cambridge University Press.

Smith, V. L. (1976). "Experimental Economics: Induced Value Theory." *American Economic Review, 66* (2), 274–79.

Smuts, B. B. (1995). "The Evolutionary Origins of Patriarchy." *Human Nature, 6* (1), 1–32.

Sokol, S. (2011). "Beit Shemesh Goes to the Streets." *Jerusalem Post,* Features Section, December 24, 2011, p. 7.

Sola, C., and Tongiorgi, P. (1996). "The Effect of Salinity on the Chemotaxis of Glass Eels, Anguilla, to Organic Earthy and Green Odorants." *Environmental Biology of Fishes, 47* (2), 213–18. doi: 10.1007/BF00005045

Solove, D. J. (2007). *The Future of Reputation: Gossip, Rumor, and Privacy on the Internet.* New Haven: Yale University Press.

Somit, A., and Peterson, S. A. (1997). *Darwinism, Dominance, and Democracy: The Biological Bases of Authoritarianism.* Westport, Conn.: Praeger.

Sorabji, C. (2006). "Managing Memories in Post-War Sarajevo: Individuals, Bad Memories, and New Wars." *Journal of the Royal Anthropological Institute, 12* (1), 1–18. doi: 10.1111/j.1467-9655.2006.00278.x

Sowell, T. (2007). *A Conflict of Visions: Ideological Origins of Political Struggles.* New York: Basic Books.

———. (2011). *Economic Facts and Fallacies* (2nd ed.). New York: Basic Books.

Spelke, E. S. (1990). "Principles of Object Perception." *Cognitive Science, 14,* 29–56.

———. (2000). "Core Knowledge." *American Psychologist, 55* (11), 1233–43.

Spelke, E. S., and Kinzler, K. D. (2007). "Core Knowledge." *Developmental Science, 10* (1), 89–96.

Sperber, D. (1997). "Intuitive and Reflective Beliefs." *Mind and Language, 12* (1), 17.

——. (2000a). "Metarepresentation in an Evolutionary Perspective." In D. Sperber (ed.), *Metarepresentations: A Multidisciplinary Perspective,* pp. 3–16. Oxford: Oxford University Press.

——. (2000b). "An Objection to the Memetic Approach to Culture." In R. Aunger (ed.), *Darwinizing Culture: The Status of Memetics as a Science,* pp. 163–73. Oxford: Oxford University Press.

——. (2002). "In Defense of Massive Modularity." In E. Dupoux (ed.), *Language, Brain and Cognitive Development: Essays in Honor of Jacques Mehler,* pp. 47–57. Cambridge, Mass.: MIT Press.

Sperber, D., and Baumard, N. (2012). "Moral Reputation: An Evolutionary and Cognitive Perspective." *Mind and Language, 27* (5), 495–518. doi: 10.1111/mila.12000

Sperber, D., and Claidière, N. (2006). "Why Modeling Cultural Evolution Is Still Such a Challenge." *Biological Theory, 1* (1), 20–22.

Sperber, D., Clément, F., Heintz, C., Mascaro, O., Mercier, H., Origgi, G., and Wilson, D. (2010). "Epistemic Vigilance." *Mind and Language, 25* (4), 359–93. doi: 10.1111/j.1468-0017.2010.01394.x

Sperber, D., and Wilson, D. (1986). *Relevance: Communication and Cognition.* New York: Academic Press.

——. (1995). *Relevance: Communication and Cognition* (2nd ed.). Oxford: Blackwell.

Sprecher, S., Sullivan, Q., and Hatfield, E. (1994). "Mate Selection Preferences: Gender Differences Examined in a National Sample." *Journal of Personality and Social Psychology, 66* (6), 1074–80. doi: 10.1037/0022-3514.66.6.1074

Stark, R. (2003). "Upper Class Asceticism: Social Origins of Ascetic Movements and Medieval Saints." *Review of Religious Research, 41,* 5–19.

Stearns, S. C. (1992). *The Evolution of Life Histories.* Oxford: Oxford University Press.

Stearns, S. C., Allal, N., and Mace, R. (2008). "Life History Theory and Human Development." In C. Crawford, D. Krebs, C. Crawford, and D. Krebs (eds.), *Foundations of Evolutionary Psychology,* pp. 47–69. New York: Taylor and Francis / Lawrence Erlbaum.

Stepanoff, C. (2014). *Chamanisme, rituel et cognition chez les Touvas de Sibérie du Sud.* Paris: Éditions de la Maison des Sciences de l'Homme.

Stewart, J. J. (2014). "Muslim-Buddhist Conflict in Contemporary Sri Lanka." *South Asia Research, 34* (3), 241–60.

Stocking, G. W. (1984). *Functionalism Historicized: Essays on British Social Anthropology.* Madison: University of Wisconsin Press.

Stubbersfield, J. M., Tehrani, J. J., and Flynn, E. G. (2014). "Serial Killers, Spiders and Cybersex: Social and Survival Information Bias in the Transmission of Urban Legends." *British Journal of Psychology, 106* (2), 288–307. doi: 10.1111/bjop.12073

Sugiyama, L. (1996). "In Search of the Adapted Mind: Cross-Cultural Evidence for Human Cognitive Adaptations among the Shiwiar of Ecuador and the Yora of Peru." Ph.D. dissertation, University of California, Santa Barbara.

Sulikowski, U. (1993). "Eating the Flesh, Eating the Soul: Reflections on Politics, Sorcery and Vodun in Contemporary Benin." In J.-P. Chrétien (ed.), *L'invention religieuse en Afrique: Histoire et religion en Afrique noire,* pp. 387–402. Paris: Karthala.

Surian, L., Caldi, S., and Sperber, D. (2007). "Attribution of Beliefs by 13-Month-Old Infants." *Psychological Science, 18* (7), 580–86.

Sussman, N. M., and Rosenfeld, H. M. (1982). "Influence of Culture, Language, and Sex on Conversational Distance." *Journal of Personality and Social Psychology, 42* (1), 66–74. doi: 10.1037/0022-3514.42.1.66

Symons, D. (1979). *The Evolution of Human Sexuality.* New York: Oxford University Press.

———. (1992). "On the Use and Misuse of Darwinism in the Study of Human Behavior." In J. H. Barkow, L. Cosmides, and J. Tooby (eds.), *The Adapted Mind: Evolutionary Psychology and the Generation of Culture,* pp. 137–59. New York: Oxford University Press.

Szechtman, H., and Woody, E. (2004). "Obsessive-Compulsive Disorder as a Disturbance of Security Motivation." *Psychological Review, 111* (1), 111–27.

Tajfel, H. (1970). "Experiments in Inter-Group Discrimination." *Scientific American, 223,* 96–102.

Tajfel, H., Billig, M., and Bundy, R. (1971). "Social Categorization and Intergroup Behaviour." *European Journal of Social Psychology, 1,* 149–78.

Talmy, L. (1988). "Force Dynamics in Language and Cognition." *Cognitive Science, 12* (1), 49–100. doi: http://dx.doi.org/10.1016/0364-0213(88)90008-0

———. (2000). *Toward a Cognitive Semantics.* Cambridge, Mass.: MIT Press.

Tambiah, S. J. (1992). *Buddhism Betrayed? Religion, Politics, and Violence in Sri Lanka.* Chicago: University of Chicago Press.

Tarde, G. (1903). *The Laws of Imitation.* New York: Henry Holt.

Taves, A. (2009). *Religious Experience Reconsidered: A Building-Block Approach to the Study of Religion and Other Special Things.* Princeton: Princeton University Press.

Taylor, C. C. (1999). *Sacrifice as Terror: The Rwandan Genocide of 1994.* Oxford: New York.

Tedeschi, J. T., Schlenker, B. R., and Bonoma, T. V. (1971). "Cognitive Dissonance: Private Ratiocination or Public Spectacle?" *American Psychologist, 26* (8), 685–95. doi: 10.1037/h0032110

Tetlock, P. E., and Goldgeier, J. M. (2000). "Human Nature and World Politics: Cognition, Identity, and Influence." *International Journal of Psychology, 35* (2), 87–96. doi: 10.1080/002075900399376

Thomas, K. (1997). *Religion and the Decline of Magic: Studies in Popular Beliefs in Sixteenth and Seventeenth Century England.* New York: Oxford University Press.

Thomas, S. A. (2007). "Lies, Damn Lies, and Rumors: An Analysis of Collective Efficacy, Rumors, and Fear in the Wake of Katrina." *Sociological Spectrum, 27* (6), 679–703. doi: 10.1080/02732170701534200

Tomasello, M. (2008). *Origins of Human Communication.* Cambridge, Mass.: Bradford Books and MIT Press.

———. (2009). *Why We Cooperate.* Cambridge, Mass.: MIT Press.

Tooby, J. (1982). "Pathogens, Polymorphism, and the Evolution of Sex." *Journal of Theoretical Biology, 97* (4), 557–76. doi: http://dx.doi.org/10.1016/0022-5193(82)90358-7

Tooby, J., and Cosmides, L. (1988). "The Evolution of War and Its Cognitive Foundations." In J. Tooby and L. Cosmides (eds.), *Evolutionary Psychology: Foundational Papers.* Cambridge, Mass.: MIT Press.

———. (1992). "The Psychological Foundations of Culture." In J. H. Barkow, L. Cosmides, et al. (eds.), *The Adapted Mind: Evolutionary Psychology and the Generation of Culture,* pp. 19–136. New York: Oxford University Press.

———. (1995). "Mapping the Evolved Functional Organization of Mind and Brain." In M. S. Gazzaniga et al. (eds.), *The Cognitive Neurosciences,* pp. 1185–97. Cambridge, Mass.: MIT Press.

———. (2010). "Groups in Mind: The Coalitional Roots of War and Morality." In H. Høgh-Olesen (ed.), *Human Morality and Sociality: Evolutionary and Comparative Perspectives,* pp. 191–234. New York: Palgrave MacMillan.

——— (eds.). (2005). *Conceptual Foundations of Evolutionary Psychology.* Hoboken, N.J.: John Wiley and Sons.

Tooby, J., and DeVore, I. (1987). "The Reconstruction of Hominid Behavioral Evolution through Strategic Modeling." In W. Kinzey (ed.), *Primate Models of Hominid Behavior,* pp. 183–237. New York: SUNY Press.

Trigger, B. G. (2003). *Understanding Early Civilizations: A Comparative Study.* Cambridge: Cambridge University Press.

Tuomela, R. (2013). *Social Ontology: Collective Intentionality and Group Agents.* New York: Oxford University Press.

Turchin, P. (2007). *War and Peace and War: The Rise and Fall of Empires.* London: Penguin.

Turiel, E. (1983). *The Development of Social Knowledge: Morality and Convention.* Cambridge: Cambridge University Press.

Valeri, V. (1985). *Kingship and Sacrifice: Ritual and Society in Ancient Hawaii.* Chicago: University of Chicago Press.

van Dijk, E., and Wilke, H. (1997). "Is It Mine or Is It Ours? Framing Property Rights and Decision Making in Social Dilemmas." *Organizational Behavior and Human Decision Processes, 71* (2), 195–209. doi: 10.1006/obhd.1997.2718

Van Schaik, C. P., and Van Hooff, J. A. (1983). "On the Ultimate Causes of Primate Social Systems." *Behaviour, 85* (1–2), 91–117.

van Vugt, M. (2006). "Evolutionary Origins of Leadership and Followership." *Personality and Social Psychology Review, 10* (4), 354–71. doi: 10.1207/s15327957pspr1004_5

van Vugt, M., Cremer, D. D., and Janssen, D. P. (2007). "Gender Differences in Cooperation and Competition: The Male-Warrior Hypothesis." *Psychological Science, 18* (1), 19–23. doi: 10.1111/j.1467-9280.2007.01842.x

Viding, E., and Larsson, H. (2010). "Genetics of Child and Adolescent Psychopathy." In R. T. Salekin, D. R. Lynam, R. T. Salekin, and D. R. Lynam (eds.), *Handbook of Child and Adolescent Psychopathy,* pp. 113–34. New York: Guilford Press.

Waldron, M., Heath, A. C., Turkheimer, E., Emery, R., Bucholz, K. K., Madden, P. A. F., and Martin, N. G. (2007). "Age at First Sexual Intercourse and Teenage Pregnancy in Australian Female Twins." *Twin Research and Human Genetics, 10* (3), 440–49. doi: 10.1375/twin.10.3.440

Walker, R. S., Hill, K. R., Flinn, M. V., and Ellsworth, R. M. (2011). "Evolutionary History of Hunter-Gatherer Marriage Practices." *PLoS One, 6* (4), e19066. doi: 10.1371/journal.pone.0019066

Ward, T. B. (1994). "Structured Imagination: The Role of Category Structure in Exemplar Generation." *Cognitive Psychology, 27* (1), 1–40.

———. (1995). "What's Old about New Ideas?" In S. M. Smith, T. B. Ward, et al. (eds.), *The Creative Cognition Approach,* pp. 157–78. Cambridge, Mass.: MIT Press.

Ward-Perkins, B. (2005). *The Fall of Rome: And the End of Civilization.* Oxford: Oxford University Press.

Washburn, S., and Lancaster, C. (1968). "The Evolution of Hunting." In R. Lee and I. DeVore (eds.), *Man the Hunter,* pp. 293–303. Chicago: Aldine Press.

Wellmann, H., and Estes, D. (1986). "Early Understandings of Mental Entities: A Re-examination of Childhood Realism." *Child Development, 57,* 910–23.

Werker, J. F., and Tees, R. C. (1999). "Influences on Infant Speech Processing: Toward a New Synthesis." *Annual Review of Psychology, 50* (1), 509–35. doi: 10.1146/annurev.psych.50.1.509

Wertsch, J. V. (2002). *Voices of Collective Remembering.* New York: Cambridge University Press.

Westermarck, E. (1921). *The History of Human Marriage,* vol. 5. London: Macmillan.

Wetzel, C. G., and Walton, M. D. (1985). "Developing Biased Social Judgments: The False-Consensus Effect." *Journal of Personality and Social Psychology, 49* (5), 1352–59.

Whitehouse, H. (1992). "Memorable Religions: Transmission, Codification and Change in Divergent Melanesian contexts." *Man, 27,* 777–97.

———. (1995). *Inside the Cult: Religious Innovation and Transmission in Papua New Guinea.* Oxford: Oxford University Press.

———. (2000). *Arguments and Icons: Divergent Modes of Religiosity.* Oxford: Oxford University Press.

———. (2004). *Modes of Religiosity.* Walnut Creek, Calif.: Altamira Press.

Whiten, A., Custance, D. M., Gomez, J.-C., Teixidor, P., and Bard, K. A. (1996). "Imitative Learning of Artificial Fruit Processing in Children (Homo sapiens) and Chimpanzees (Pan troglodytes)." *Journal of Comparative Psychology, 110* (1), 3.

Whitson, J. A., and Galinsky, A. D. (2008). "Lacking Control Increases Illusory Pattern Perception." *Science, 322* (5898), 115.

Wiessner, P. (2005). "Norm Enforcement among the Ju/'hoansi Bushmen: A Case of Strong Reciprocity?" *Human Nature, 16* (2), 115–45.

Wilkinson, J. F. (1891). *The Friendly Society Movement: Its Origin, Rise, and Growth: Its Social, Moral, and Educational Influences.* London: Longmans, Green.

Williams, D. R., and Mohammed, S. A. (2009). "Discrimination and Racial Disparities in Health: Evidence and Needed Research." *Journal of Behavioral Medicine, 32,* 20–47.

Wilson, E. O. (1975). *Sociobiology: The New Synthesis.* Cambridge, Mass.: Belknap Press of Harvard University Press.

———. (1998). *Consilience: The Unity of Knowledge.* London: Little, Brown.

———. (1999). *Consilience: The Unity of Knowledge.* New York: Vintage.

Wilson, M., and Daly, M. (1992). "The Man Who Mistook His Wife for a Chattel." In J. H. Barkow and L. Cosmides (eds.), *The Adapted Mind: Evolutionary Psychology and the Generation of Culture,* pp. 289–322. London: Oxford University Press.

———. (1997). "Life Expectancy, Economic Inequality, Homicide, and Reproductive Timing in Chicago Neighbourhoods." *British Medical Journal, 314* (7089), 1271.

———. (1998). "Lethal and Nonlethal Violence against Wives and the Evolutionary Psychology of Male Sexual Proprietariness." In R. E. Dobash and R. P. Dobash (eds.), *Rethinking Violence against Women,* pp. 199–230. Thousand Oaks, Calif.: Sage.

Wolf, A. P. (1995). *Sexual Attraction and Childhood Association: A Chinese Brief for Edward Westermarck.* Stanford, Calif.: Stanford University Press.

Wood, G. E. (2002). *Fifty Economic Fallacies Exposed.* London: Institute of Economic Affairs.

Woodward, A. L. (2003). "Infants' Developing Understanding of the Link between Looker and Object." *Developmental Science, 6* (3), 297–311.

Woody, E., and Szechtman, H. (2011). "Adaptation to Potential Threat: The Evolution, Neurobiology, and Psychopathology." *Neuroscience and Biobehavioral Reviews, 35* (4), 1019–33.

Worstall, T. (2014). *20 Economics Fallacies.* N.p.: Searching Finance.

Wrangham, R. W., Jones, J. H., Laden, G., Pilbeam, D., and Conklin-Brittain, N. (1999). "The Raw and the Stolen: Cooking and the Ecology of Human Origins." *Current Anthropology, 40* (5), 567–94. doi: 10.1086/300083

Wrangham, R. W., and Peterson, D. (1997). *Demonic Males: Apes and the Origins of Violence.* London: Bloomsbury.

Xenophon. (1960). *Cyropaedia.* Cambridge, Mass.: Harvard University Press.

Xiaoxiaosheng, L.-L. (1993). *The Plum in the Golden Vase.* Trans. D. T. Roy. Princeton: Princeton University Press.

Xu, F., and Tenenbaum, J. B. (2007). "Word Learning as Bayesian Inference." *Psychological Review, 114* (2), 245.

Yamagishi, T., and Mifune, N. (2009). "Social Exchange and Solidarity: In-Group Love or Out-Group Hate?" *Evolution and Human Behavior, 30* (4), 229–37.

Yan, Y. (1996). *The Flow of Gifts: Reciprocity and Social Networks in a Chinese Village.* Stanford, Calif.: Stanford University Press.

Zahavi, A., and Zahavi, A. (1997). *The Handicap Principle: A Missing Piece of Darwin's Puzzle.* New York: Oxford University Press.

Zivi, K. (2014). "Performing the Nation: Contesting Same-Sex Marriage Rights in the United States." *Journal of Human Rights, 13* (3), 290–306. doi: 10.1080/14754835.2014.919216

Acknowledgments

As it took a long time for this book to take its final shape, I have incurred many debts along the way. I was able to sketch the first elements during a sabbatical year made possible by a fellowship from the John S. Guggenheim Foundation. In the following years, I must have taken inspiration from many individuals and their published work, many times more than is credited in bibliographic endnotes and explicit discussion. Beyond these, I must express special thanks to a few individuals whose contribution was of special import. I would be remiss if I did not acknowledge my great intellectual debt to Dan Sperber, Leda Cosmides, and John Tooby, incurred over years of friendship, argument, and conversation—exchanges that certainly left the ledger grossly unbalanced. I also owe many of my better ideas to constant intellectual exchanges with Pierre Lienard, as well as many discussions of possible research programs with Nicolas Baumard. Finally, I am very grateful to friends who had the patience to read through drafts of the book, to comment on their many flaws, and to suggest appropriate rhetorical or substantive remedies—so, very special thanks to Nicolas Baumard, Coralie Chevallier, Leda Cosmides, Pierre Lienard, Robert McCauley, Hugo Mercier, Olivier Morin, Michael Bang Petersen, and Jim Wertsch.

Index